WITHDRAWN

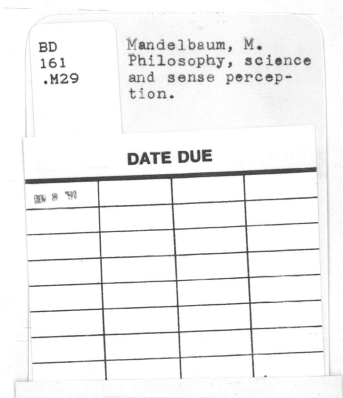

PHILOSOPHY, SCIENCE, AND SENSE PERCEPTION:

Historical and Critical Studies

PHILOSOPHY
SCIENCE AND
SENSE PERCEPTION:

Historical and Critical Studies

By

MAURICE MANDELBAUM

The Johns Hopkins Press, Baltimore

Copyright © 1964 by The Johns Hopkins Press
Baltimore, Maryland 21218
Printed in the United States of America
Library of Congress Catalog Card Number 64–16312

Originally published, 1964
Second printing, 1966

Johns Hopkins Paperbacks edition, 1966

This book has been brought to publication
with the assistance of a grant from
The Ford Foundation.

To

A. M. M.

in gratitude

PREFACE

THE FOUR ESSAYS WHICH constitute this book have not previously been published. While each is independent of the others, and the argument of each is therefore to be judged on its own merits, there is at least one theme common to them: the question of how one might hope to establish or to defend a critical realism. Of late, no form of critical realism has seemed to commend itself to those who have been concerned either with epistemology in general or with the problem of perception in particular; therefore, if this position is to receive reexamination it may be helpful to view it in historical perspective. However, rather than choose to regard it from the nearer distance of the debates among realists in the first decades of the twentieth century, I have gone back to an earlier point in its history: the seventeenth century.

It is in its seventeenth-century formulations, and in eighteenth-century attacks upon it, that we can, I think, best appreciate one of the most usual and fundamental characteristics of critical realism: its acceptance of the belief that scientific inquiries are directly relevant to epistemological issues. And, in my opinion, it is in no small measure due to the fact that contemporary philosophers tend to draw a sharp distinction between scientific and philosophic problems that critical realism has suffered an almost total eclipse in our time.

If one thinks back on the seventeenth century, and if one takes Galileo, Descartes, Spinoza, Boyle, or Locke as examples, it should be evident that what was considered to be a philosophic theory of sense perception was not construed as being independent of empirical investigations regarding the nature and structures of material objects, the principles of optics, or the human constitution. Such is no longer the case. Instead, following the lead of Berkeley and of Hume, a

vii

philosophical theory of sense perception is today either considered to be basic to the interpretation of the results of empirical investigations, or the two sorts of inquiry, the philosophical and the scientific, are viewed as moving in separate realms of discourse, each one being concerned with different problems, and with no points of contact existing between them.

It would be quixotic to think that one could change this climate of opinion merely by putting forward still another philosophical theory of perception. However, if it can be established that others have in the past held a view opposed to the one which now dominates our thought, and if at the same time it can be shown that at least some of the arguments for the current separation of epistemology and scientific inquiry are more vulnerable than they are sometimes suspected of being, it may be hoped that philosophers will reconsider their present assumptions.

It is, then, in this hope that the present, relatively independent studies have been brought together in one volume. The first of them is primarily historical in intention: it aims to offer a reinterpretation of certain of Locke's views through relating his epistemology to the science of his day, instead of viewing his critical realism in the light of what Berkeley, operating on the basis of quite different assumptions, found to be objectionable in it. The second study is closely related to the first in that it seeks to show how Boyle and Newton attempted to vindicate the sort of realism which Locke accepted from them, and which was basic to their science no less than it was to his epistemology. The third study is not to an equal degree historical. Rather, it is concerned with Hume primarily as an example of the philosophic revolt against the epistemological convictions of the seventeenth century, and against the critical realism which they entailed. In this essay there is an attempt to show that Hume's phenomenalism is untenable, although it is admitted that should one wish to accept it without adducing arguments in its favor one could consistently maintain it. However, in the final essay a more positive conclusion is suggested. The arguments used to cast doubt upon Hume's solution of the epistemological problem are shown not to lead to the direct realism which one identifies with the position of either G. E. Moore or Gilbert Ryle. Through a criticism of their positions it is argued that the conclusions reached in scientific inquiries are in fact relevant to the truth or falsity of epistemological views. It is then suggested that when we sift and assess our ordinary perceptual experience, and when we trace the cumulative achievements of physics, physiology, and experimental psychology, noting how these relate to problems of perception, the time is at hand when we can reassert and defend a radical form of critical realism.

MAURICE MANDELBAUM

The Johns Hopkins University
September, 1963.

CONTENTS

PHILOSOPHY, SCIENCE, AND SENSE PERCEPTION:

Historical and Critical Studies

1

LOCKE'S REALISM

THE AIM OF THIS essay is to untangle a badly snarled set of problems in Locke's philosophy by picking up a loose end which has not usually been thought to lead into the heart of the tangle. This loose end is to be found in the fact that Locke, like Boyle and Newton, was an atomist. As we shall later see, neither Boyle nor Newton regarded atomism as a speculative or metaphysical system, but as an inductively confirmable theory basic to their new experimental philosophy. Therefore, when one recalls Locke's faith in the achievements of the new science, and his high opinion of Boyle and of Newton, it is surely not surprising to find him taking the truth of atomism for granted. This, however, entailed the acceptance of a point which was to be called into question by some of his successors: unlike them, Locke was never led to doubt the existence of an independent world of physical objects. Neither did he doubt that this world actually possessed those characteristics which the new experimental science attributed to it. Furthermore, throughout his analysis of human knowledge, he viewed our experience as taking its rise from the action of physical objects upon us. To be sure, in the fourth book of the *Essay*,[1] Locke did raise the question of how we can justify our belief in a world of objects lying outside of our experience; however, neither here nor elsewhere did he challenge the truth of that belief.[2] On the contrary, his realism was, I suggest,

---

[1] Full references to the works and editions cited will be found in the Bibliography. In those cases in which it is likely to be useful to do so, I shall give chapter and section references, followed by a page reference to the edition which I have used. In this case the reference is to *An Essay Concerning Human Understanding*, Bk. IV, Ch. XI (II, 325 ff.).

[2] In fact, his language shows genuine impatience with such challenges.

an assumption which provided a framework within which his whole account of our knowledge was set. A passage which illustrates this fact, and which is particularly telling because of its position in the long history of the writing of the *Essay,* is the opening paragraph of Draft A of the *Essay:*

> I imagin that all knowledg is founded on and ultimately derives its self from sense, or something analogous to it and may be calld sensation which is donne by our senses conversant about particular objects which give us the simple Ideas or Images of things and thus we come to have Ideas of heat and light, hard and soft which are noe thing but the reviveing again in our mindes those imaginations which those objects when they affected our senses caused in us whether by motion or otherwise it matters not here to consider, and thus we doe when we conceive heat or light, yellow or blew, sweet or bitter &c.[3]

Looking back upon such a statement in the light of Berkeley's critique of Locke, we may of course feel that Locke is in a hopeless

---

For example, in the chapter just cited, he says: " I think nobody can, in earnest, be so sceptical as to be uncertain of the existence of those things which he sees and feels. At least, he that can doubt so far, (whatever he may have with his own thoughts,) will never have any controversy with me; since he can never be sure I say anything contrary to his own opinion " (Sec. 3 [II, 327]). And in Sec. 8 (II, 332) he says:

> But yet, if after all this any one will be so sceptical as to distrust his senses, and to affirm that all we see and hear, feel and taste, think and do, during our whole being, is but the series and deluding appearances of a long dream, whereof there is no reality; and therefore will question the existence of all things, or our knowledge of anything: I must desire him to consider, that, if all be a dream, then he doth but dream that he makes the question, and so it is not much matter that a waking man should answer him.

[3] In even the last edition of the *Essay,* when Locke raised the question of how we might justify our belief in the independent existence of physical objects, he used the same framework to explain the origins of human knowledge. For example, he says: " No particular man can know the existence of any other being, but only when, by actual operating upon him, it makes itself perceived by him." (Bk. IV, Ch. XI, Sec. 1 [II, 325]): also, he says: " It is therefore the *actual receiving* of ideas from without that gives us notice of the existence of other things, and makes us know, that something doth exist at that time without us, which causes that idea in us " (*ibid.,* Sec. 2).

muddle. However, it is usually misleading to read intellectual history backwards, and in this case it is especially so. Berkeley challenged an important tradition which Locke had consciously and willingly accepted: the tradition of those who, like Boyle, looked upon scientific inquiry as an essential basis for sound philosophizing. In opposition to this tradition, Berkeley wished to circumscribe the philosophic import of the conclusions which were currently being drawn from experimental inquiries, and throughout his works he sought to free philosophic questions from any direct dependence upon science. Consequently, the more important the place which we must assign to Boyle, and to other representatives of the new science, in the formation of Locke's thought, the more misleading will it be to interpret Locke in the light of Berkeley's criticisms of him.

This point is of sufficient importance to bear further emphasis. I think it will be conceded that any approach to Locke through Berkeley's criticisms will be bound to stress difficulties in his doctrine of primary and secondary qualities, and in his doctrine of material substances. However, these are two points at which a wholehearted acceptance of atomism commits one to positions which are in all respects opposed to Berkeleian views. Therefore, if Locke did in fact accept atomism as a scientifically established theory, and if he also accepted science as a basis for a theory of knowledge, a fair interpretation of his actual views would have to approach his position by reading the *Essay* in the light of his atomism, and not merely as an epistemological treatise devoid of a scientific substructure. This, however, means that one should interpret his theory of knowledge in the light of his relations to Boyle, rather than merely treating him as a forerunner of Berkeley.[4] To be sure, even when we look at Locke in these terms not all of the ambiguities and

---

[4] In this connection it must be remembered that Boyle was looked upon as the chief proponent of atomism, and that Berkeley came to be its chief philosophic opponent.

The close personal relations between Locke and Boyle are well known. Locke became acquainted with Boyle not later than the early 1660's, and their contacts were practically uninterrupted. Boyle, who died in 1691, named Locke as one of his scientific and literary executors. As to Boyle's place in the opinion of Locke and his friends, it is surely not irrelevant that Syden-

inconsistencies in his thought will disappear. Under no circumstances can he be counted among the clearest and most consistent of philosophers. However, I hope that by looking at him in this way we shall no longer have to attribute to him that degree of obtuseness which epistemologists who favor a Berkeleian view of the relations between philosophy and science have been pleased to attribute to him.

First, however, it will be necessary briefly to document the fact that Locke was an atomist, for this has sometimes been challenged.

I

As one example among many to illustrate Locke's unquestioning acceptance of atomism, we may cite the following passage:

My present purpose being only to inquire into the knowledge the mind has of things, by those ideas and appearances which God has fitted it to receive from them, and how the mind comes by that knowledge; rather than into their causes or manner of production, I shall not, contrary to the design of this *Essay*, set myself to inquire philosophically into the peculiar constitution of *bodies*, and the configuration of parts, whereby *they* have the power to produce in us the ideas of their sensible qualities. I

---

ham dedicated the first two editions of his *Medical Observations Concerning the History and Cure of Acute Diseases* to Boyle.

Aaron and Gibson acknowledge the great influence of Boyle upon Locke, even though they do not explore it with care. (Aaron, for example, stresses the influence of Gassendi rather than of Boyle on Locke's atomism.) Ollian treats it at greater length, but less discriminatingly. O'Connor fails to mention Boyle. Perhaps the most careful as well as the most suggestive treatment of their philosophical relationship is to be found in Anderson, " The Influence of Contemporary Science on Locke's Method and Results." The reader will note, however, that my views diverge rather widely from those of Anderson.

The other study which most clearly shows Locke's affinity to Boyle is C. Baeumker, " Ueber die Lockesche Lehre von den primären und sekundären Qualitäten." In a doctoral dissertation entitled " John Locke und die mechanische Naturauffassung," W. Schröder also noted that Locke's assumption of the truth of atomistic mechanism separated him from Berkeley and from Hume (cf. pp. 46–49).

shall not enter any further into that disquisition; it sufficing to my purpose to observe, that gold or saffron has a power to produce in us the idea of yellow, and snow or milk, the idea of white, which we can only have by our sight; without examining the texture of the parts of those bodies, or the particular figures or motion of the particles which rebound from them, to cause in us that particular sensation: though, when we go beyond the bare ideas in our minds, and would inquire into their causes, we cannot conceive anything else to be in any sensible object, whereby it produces different ideas in us, but the different bulk, figure, number, texture, and motion of its insensible parts.[5]

Passages of this sort are to be found throughout the *Essay*,[6] and they are to be found in its earlier drafts as well.[7] It is surely also relevant that Leibniz classed Locke as belonging to the party of Gassendi.[8]

We must, however, note that it is sometimes held that even though Locke did accept the truth of atomism, he always remained skeptical of its attempts to explain the particular phenomena of nature. If this were true, such a skeptical reserve would separate him from the tradition of Boyle and of other seventeenth-century scientists, and would weaken the thesis which I here wish to support; it is therefore necessary to face this challenge immediately.

A number of different lines of evidence have been brought forward to suggest that Locke was skeptical of the usefulness of atomism as an explanation of events in nature. One of these has been found in his medical fragments, and another in his discussion of natural philosophy in *Some Thoughts Concerning Education*; however, the weight of the evidence must of course be found in the *Essay*, and this evidence has been marshalled by R. M. Yost, Jr., in an interest-

---

[5] *Essay*, Bk. II, Ch. XXI, Sec. 75 (I, 373–74).

[6] For example, in Book II they are to be found in Ch. IV, Sec. 4; Ch. VIII, Secs. 4, 11, 13–17; Ch. XXIII, Sec. 11; Ch. XXXI, Sec. 6. In Book III, see Ch. VI, Sec. 6; in Book IV, Ch. III, Secs. 16 and 25, and Ch. X, Sec. 10 also illustrate Locke's atomism.

(The page references in the Fraser edition are as follows: I, 154–55, 167, 171, 172–74, 401, 506–8; II, 61–62, 205–6, 216–17, 314.)

[7] For example, *Draft B*, pp. 198–99 and 209. *Draft C* is also quite explicitly atomistic, cf. Aaron, *John Locke*, p. 68.

[8] *New Essays Concerning Human Understanding*, I, i (Langley ed., 65).

ing article which merits careful consideration.[9] I shall deal with each of these lines of evidence in turn.

Among the medical fragments the one most frequently cited as a means of establishing Locke's skepticism regarding the usefulness of the corpuscular theory is to be found in his fragment on anatomy.[10] The relevant portion of that fragment is summarized by Bourne in the following way. (I shall italicize those portions of the summary which are in Locke's own words.)

> "*Anatomy, no question, is absolutely necessary to a surgeon, and to a physician who would direct a surgeon in incision, trepanning, and several other operations.*" Locke pointed out other cases in which anatomy is useful, if not necessary, to medical practice. Then he propounded what every one now-a-days must regard as a strange heresy for such a man to hold. "*But that anatomy,*" he said, "*is like to afford any great improvements to the practice of physic, or assist a man in the finding out and establishing a true method, I have reason to doubt. All that anatomy can do is only to show us the gross and sensible parts of the body, or the vapid and dead juices, all which, after the most diligent search, will be no more able to direct a physician how to cure a disease than how to make a man; for to remedy the defects of a part whose organical constitution, and that texture whereby it operates, he cannot possibly know, is alike hard as to make a part he knows not how is made.*"[11]

It seems to me clear that in this passage Locke is not in any way challenging the usefulness of the corpuscular philosophy; he is only challenging the usefulness of gross anatomy in medicine. (Although it is to be noted that he does not deny its usefulness in surgery.) Clearly, however, descriptive anatomy is something quite different

[9] "Locke's Rejection of Hypotheses about Sub-Microscopic Events."

[10] The most complete single discussion of the medical fragments is given in Bourne, *The Life of John Locke*, I, 222 ff., although it is no longer up to date. For a summary of the known extant materials, cf. Romanell, " Grant No. 2227 . . ." as cited in the bibliography. For a recent re-editing of Locke's "De Arte Medica" fragment, cf. Gibson, *The Physician's Art*.

[11] Bourne, *The Life of John Locke*, I, 228 f. This passage is cited in connection with Locke's views on atomism by Ollian, *Le philosophie générale de Locke*, pp. 131–32, and by Yost, "Locke's Rejection . . . ," p. 129.

from investigations of the submicroscopic constituents of material objects, and in this passage it is precisely because gross anatomy does not penetrate beyond "the gross and sensible parts of the body," and does not reveal "the organical constitution, and that texture whereby it operates"—that is, because it does not reach the submicroscopic level of the corpuscular parts—that it fails to be useful for the physician.[12] To be sure, Locke does not say here that the corpuscular view of matter can itself be of positive use to the physician, but that question was not one which would necessarily be involved in a treatise on the uses and limits of anatomy.

In another of the medical fragments, *De Arte Medica,* which has also been cited in this connection,[13] Locke was surely not attacking the modern corpuscularians, but was only concerned to attack "hypotheses." As we shall see, that term was used by Newton and others to refer to what might best be called "metaphysical" explanations, in contradistinction to empirical hypotheses. What seems to have misled commentators in this matter is the fact that in his

[12] The words "constitution" and "texture" are reminiscent of Boyle, and it would surely be a mistake to think that Locke could write in this vein without having in mind Boyle's repeated defenses of the usefulness of the corpuscular theory for the physician. While Boyle's most explicit defense of this usefulness is to be found in a work entitled "On the Reconcileableness of Specific Medicines to the Corpuscular Philosophy" (*Works*, V), which postdated the fragment here in question, the same position was held by him in numerous earlier works, e. g., in "The Usefulness of Experimental Philosophy" (*Works*, II, 170–73). Also, in the same year as Locke's fragment "On Anatomy," one finds that Glanvill, on the basis of information furnished him by Oldenburg, seems to have been aware of Boyle's views regarding the question (cf. *Plus Ultra*, p. 105). Thus we may take this view as having been known, and undoubtedly accessible to Locke.

[13] Cf. Yost, "Locke's Rejection . . . ," p. 129. In this fragment Yost summarizes Locke's position in the following way (again I italicize Locke's words):

He rejected all attempts to advance medicine by trying to discover "*the hidden causes of distempers, . . . the secret workmanship of nature and the several imperceptible tools wherewith she wrought,*" believing these matters to be "*utterly out of reach*" of man's apprehension.

This, however, is a misleading summary, for the words quoted from Locke refer to "the learned men of former ages" who "putting all these fancies together, fashioned to themselves systems and hypotheses" (cf. Bourne, *The Life of John Locke*, I, 223). Clearly, however, this polemical tone would not have been adopted with respect to Boyle's form of the corpuscular theory.

medical writings Locke continually insisted upon the "historical" method, and this has not unnaturally been identified with the method of Sydenham: a method which presumably did not seek explanations, even in the form of empirical hypotheses, but confined itself to rules of practice based on past observation.[14] However, Locke's use of the term "histories" may also have been influenced by Boyle, for whom "a history" did not stand opposed to an experimental inquiry, but was a means of reaching or testing an empirical hypothesis.[15] It was in fact through his "histories" that Boyle sought to establish the applicability of the corpuscular hypothesis in case after case. And it is also to be noted that Boyle, although the staunchest defender of the usefulness of the corpuscular philosophy, believed in the usefulness of nonatomistic explanations as well.[16] Therefore, Locke's emphasis on the utility of Sydenham's method does not suggest that he disbelieved in either the truth or the usefulness of the corpuscular philosophy. At most it proves that he did not believe that it was as yet of use to the practicing physician.[17]

[14] Cf. Yost, "Locke's Rejection . . . ," p. 129, and Romanell, "Locke and Sydenham," especially pp. 315–17. Romanell's interpretation of Locke is somewhat similar to that of Yost, but is based on less evidence and wholly overlooks the extent to which Locke was an atomist. Furthermore, Romanell's attempt to find the source of the *Essay* in Locke's medical interests is unconvincing. (For other criticisms of this article, cf. Cowan, "Comments on Dr. Romanell's Article.")

[15] This seems to be overlooked by Yost, "Locke's Rejection . . . ," p. 127, n. 33. On Boyle's use of the term "histories," cf. below, p. 96, n. 66.

I do not wish to suggest that Locke's use of the term "historical" is wholly dependent on Boyle. In fact, it seems to me to be simply a term used in opposition to speculation and metaphysical reasoning, and not one which designated a specific procedural method at all. For example, when Locke says that in the *Essay* he will follow "the historical, plain method" (Intro., Sec. 2, [I, 27]), it seems hard to believe that he thought he was applying either Sydenham's method or Boyle's method to the problem of analyzing the origins, certainty, and extent of human knowledge. What he obviously did believe he was doing was giving a careful analytic account of these matters, and careful analysis was a feature which was common to the methods advocated by both Sydenham and Boyle. For an example of his use of "history" in this sense, cf. *Essay*, Bk. III, Ch. XI, Sec. 24 (II, 161).

[16] Cf. Boyle, *Works*, I, 308.

[17] It is to be noted that Boyle himself defended its usefulness in a programmatic way: he did not claim that the corpuscular philosophy could immediately be applied to the treatment of disease. Cf. the Prefatory Letter to "On the

The second major passage upon which a denial of Locke's trust in the new corpuscular philosophy might be based is to be found in his discussion of natural philosophy in sections 193 and 194 of *Some Thoughts Concerning Education*. In this passage Locke contrasts systems of natural philosophy with science, and finds all such systems wanting, though he admits that " the modern corpuscularians talk, in most things, more intelligibly than the peripatetics, who possessed the schools before them." [18] This, of course, is small praise, especially in the light of the fact that Locke only commends the study of systems of natural philosophy as a means of being able to understand the concepts frequently employed in polite society. But it is to be noted that the term " the modern corpuscularians"—a term apparently coined by Boyle—included Descartes as well as the atomists. [19] In fact, since in the same passage Locke explicitly excluded Boyle's works from the systems of natural philosophy, and since these works—along with the works of Newton—are recommended to the reader in preference to such systems, it seems not farfetched to think that the modern corpuscularians whom Locke here had in mind were primarily Descartes and Gassendi, and their followers. [20] Thus, this passage can scarcely be interpreted as express-

Reconcileableness of Specific Medicines to the Corpuscular Philosophy " (*Works*, V, 74, 75).

It is also worth noting that Locke added a passage in the second edition of the *Essay* which must surely be taken as a criticism of the alchemists (Fraser's footnote to this passage is wholly misleading), and as praise of Boyle's methods. This passage (Bk. IV, Ch. III, Sec. 16) will be mentioned in another connection in note 115, below.

[18] Locke, *Works*, IX, 185.

[19] Otherwise, could Locke have said that the modern corpuscularians immediately succeeded the peripatetics? Boyle, of course, himself included Descartes, along with Gassendi, as a corpuscularian: cf. Boyle, *Works*, I, 355.

For treatments of the history of modern corpuscular views, cf. Marie Boas, " The Establishment of the Mechanical Philosophy," and Kuhn, " Robert Boyle and Structural Chemistry in the 17th Century."

[20] Locke's notebooks show his great interest in Descartes at the time he was working on the *Essay* (cf. the selections from them given by Aaron and Gibb in *Draft A*).

He could not fail to have known Gassendi's works which were available when he was at Oxford, and which Boyle frequently cited. Although Locke only once, to my knowledge, cites Gassendi's name in those of his works which are thus far available (cf. his third letter to Stillingfleet, *Works*, IV, 420),

ing doubt as to the validity, or the scientific usefulness, of atomism as
such. On the contrary, it must be interpreted as endorsing the sort of
atomism to be found in Boyle and Newton, and restricting its adverse
criticism to those systems of natural philosophy which did not conform to what Locke took to be the new experimental method.[21]

But what, finally, shall be said concerning R. M. Yost's contention
that even though Locke did in fact accept atomism he, "unlike
many scientists and philosophers of the seventeenth century, . . .
did not believe that the employment of hypotheses about sub-microscopic events would accelerate the acquisition of empirical knowledge"?[22] A full answer to this contention would involve an examination of all of the passages which Yost uses in his argument,
and a comparison of his reading of these passages with the reading
which would follow from the interpretation of Locke's position which
I wish to set forth. Needless to say, I shall not here engage in a
detailed examination of this sort.[23] Instead, I shall bring forward

Aaron has quite rightly stressed the probable influence of the Gassendists on
Locke (cf. *John Locke*, pp. 31 ff., *passim*). As is well known, Locke did have
personal contacts with Bernier, the chief expositor of Gassendi, and he may
even have lodged in Paris with another of Gassendi's popularizers, Gilles de
Launay. (On both points, cf. Lough's edition of Locke's journals, entitled
Locke's Travels in France. To what Lough states, one must however add
that the mere fact that the journals do not cite Bernier in any connection
other than that of an oriental traveler does not in the least suggest that Locke
was not fully aware of his philosophy, nor that he did not know Bernier's
Abrégé de Gassendi. In this connection one may note that in his reply to
Stillingfleet, cited above, Locke explicitly mentions the name of Bernier along
with Gassendi.)

[21] Four years after Locke's essay on education, William Wooton drew a
similar distinction between the *hypotheses* of Gassendi, Descartes and Hobbes,
and the *theories* established by the experimental method (cf. Boas, "The
Establishment of the Mechanical Philosophy," p. 487).

[22] Yost, "Locke's Rejection . . . ," p. 111. Another way in which Yost
puts his point is to say that Locke differed from these scientists and philosophers by denying that "the nature of sub-microscopic events are discoverable" (cf. p. 120, which refers back to p. 112). This, of course, is a different
and more radical point, although Yost does not distinguish between his two
statements of his argument. I shall phrase my objections to Yost's interpretation of Locke in such a way as to allow them to be applicable to either
interpretation of his thesis.

[23] The passages which Yost (*ibid.*) cites from the *Essay* in support of his
thesis are: Bk. II, Ch. XXIII, Sec. 32, and Ch. XXXI, Sec. 6; Bk. III, Ch. VI,

one consideration which seems to me to constitute fairly strong prima facie evidence against Yost's contention, and shall then suggest two points at which his interpretations seem to me to be misleading.

The prima facie evidence which I would cite against Professor Yost is the fact that in his "Epistle to the Reader," Locke spoke of Boyle, "the great Huygenius," and "the incomparable Mr. Newton" as the master builders of the age—yet all were staunch advocates of the corpuscular hypothesis, and employed it in their scientific researches.[24] Furthermore, if Yost's interpretation were correct it is difficult to understand why in all those passages in which he interprets Locke as arguing against the usefulness of the corpuscular hypothesis, Locke never once raised objections against Boyle or the other atomists. Furthermore, in all of those passages in which Locke expressed his disinclination to enter into a detailed discussion of the constitution of material objects and of their action upon us, I find no implied skepticism whatsoever concerning the adequacy of such accounts: Locke merely holds that these problems were not part of the task which he set himself.[25]

Turning now to the more specific reasons why Professor Yost's interpretation of the relevant passages seems to me to be dubious, the first point which I should wish to make is that he fails to take into account the fact that one of Locke's fundamental motives was to stress the limitations of *all* human knowledge. When we take

Secs. 8 and 9, and Ch. X, Sec. 19; Bk. IV, Ch. III, Secs. 25, 26, 29, Ch. VI, Secs. 5, 11, 13, Ch. VIII, Sec. 9, Ch. XII, Secs. 10, 11, 12, and Ch. XVI, Sec. 12.

[24] On Newton's acceptance of the corpuscular hypothesis, and on the problem of whether the usual positivistic interpretation of his philosophy of science is adequate, the reader is referred to the next chapter.

At this point I should also like to note that if there really had been a contrast in Locke's mind between Sydenham's historical method and the corpuscularian hypotheses of Boyle, Huyghens, and Newton, as Yost seems to believe that there was, it would have been strange for Locke to have linked their names in his "Epistle to the Reader."

[25] One such passage has already been quoted, cf. pp. 4 f., above. Other well-known statements in the same vein are to be found in Introduction, Sec. 2 (I, 26), in Bk. II, Ch. VIII, Sec. 22 (I, 177 f.), and Bk. II, Ch. XXI, Sec. 2 (I, 503 f.) of the *Essay*.

into account this desire to restrain the claims and pretensions of men, it is not unnatural that Locke should stress the limitations of our knowledge of the internal structure of corporeal substances. Such a stress need not then be regarded as evincing any special reserve concerning scientific inference; it would merely be one application of Locke's general contention that all of our knowledge is limited to what is suitable to our estate.[26] Furthermore, it should be noted that in these discussions a modern reader may discern a more skeptical note than was actually intended by Locke, for in Locke's terminology there is an absolute difference between what is to be denominated as "science" or "knowledge," and what was to be called "opinion" or "probability."[27] Bearing these points in mind, many of the passages cited by Yost seem to me not to express skeptical reserve concerning the corpuscular hypothesis. What I find lacking in Yost's treatment of them is an analysis of each of these passages in its context, and an attempt to determine against what or whom each was directed. Such analyses seem to me to show that Locke was attacking certain widespread human pretensions, and dogmatism, and that his opponents were not in fact the atomists of his age.[28]

[26] For example, cf. *ibid.*, Bk. II, Ch. XXIII, Sec. 12–13.

[27] While Yost ("Locke's Rejection . . .") recognizes this distinction in Locke (cf. p. 123) he does not—it seems to me—exercise sufficient care in applying it when interpreting some of the passages with which he deals. (For example, compare his use of Bk. IV, Ch. III, Sec. 26 on p. 125 f. of his article with the place of that discussion in the *Essay*.)

[28] Furthermore, in the passages quoted by Yost, Locke's opponents are sometimes "the Schoolmen." For example, in the *Essay*, Bk. III, Ch. VI, Secs. 8, 9, and 10, Locke is putting forward his own doctrine of the contrast between the nominal essences and the real essences of substances in opposition to a scholastic doctrine of species. What Sec. 9 aims to show is that we do not reach real essences through "sorting" things and "disposing them into certain classes under names," that is, through using their nominal essences. That this passage should be used by Yost as a key part of the direct evidence for his thesis seems to me to illustrate the importance of trying to determine against whom Locke is arguing. I cannot see that it can be taken as arguing against the corpuscular views of his contemporaries when it is read in connection with the preceding section (Sec. 8) on "species" and the succeeding section on "substantial forms."

Similarly, in "An Examination of P. Malebranche's Opinion of Seeing All Things in God," which was written in approximately 1694, but was only posthumously published, Locke clearly accepts an account of sense perception

This brings me to the second point at which Yost's interpretation seems to me to be in error. Throughout the *Essay* Locke is primarily concerned with our ordinary everyday knowledge, and not with the problems of scientific inference.[29] Furthermore, whether consistently or not, Locke always draws a contrast between our "sensible ideas" and "insensible corpuscles," between what is accessible to us in direct experience and the real essences of objects.[30] Professor Yost tends to intepret all such passages as expressing skepticism regarding the possibility of attaining reliable explanations of phenomena in terms of their atomic constitutions. However, when we recall that Locke is really concerning himself with our everyday experience, and not with scientific inference, these passages take on a quite different meaning: not being concerned with the problem of how we know the internal constitution of things, the accent in these passages falls on the disparity between common knowledge and the fruits of scientific inquiry.[31] What is of course lacking is an analysis of scientific inquiry, and a justification of scientific inference to "insensible corpuscles." However, if we may assume that

which is based on corpuscularianism, and he rejects both the Scholastic doctrine of "sensible species" and any nonrealistic interpretation of the physical processes involved in the action of objects on our sense organs (Sec. 9–15, to be found in *Works*, IX, 215–19).

I am pleased to find that David A. Givner, in a recent article entitled "Scientific Preconceptions in Locke's Philosophy of Language," takes the same view of Locke's corpuscularianism and its relation to real and nominal essences as I have proposed and will more fully discuss on pp. 41–46, below.

[29] Yost recognizes that "Locke never wrote a treatise or even a chapter that was devoted exclusively to the methods of science" ("Locke's Rejection . . . ," p. 120). While I would certainly have no quarrel with Yost's further statement that it is likely that Locke "thought a good deal about the methods of empirical science and had well-considered opinions concerning them," it does not follow that when, in the *Essay*, Locke is stressing the limitations of human knowledge in general, he should be taken as stressing the limitations of scientific knowledge.

[30] For example, cf. Bk. III, Ch. VI, Sec. 9, and Bk. IV, Ch. III, Sec. 25.

[31] For example, in one of the key passages (*Essay*, Bk. II, Ch. XXXI, Sec. 6) which Yost cites in favor of his interpretation (cf. "Locke's Rejection . . . ," p. 125), the context of the passage is that of "the common idea men have" of substances such as iron or gold; it is not a discussion of what we learn through scientific inquiry. (The phrase I have quoted appears two sentences before the point at which Yost's quotation begins.)

Locke did believe that his scientific contemporaries were "master-builders, whose mighty designs, in advancing the sciences, will leave lasting monuments to the admiration of posterity," he may perhaps be excused for not having challenged their assumptions. If, then, we read those passages in which he draws a contrast between common experience and science in terms of his admiration for the natural scientists of his day, rather than in terms of Berkeley's challenge to these same scientists, Yost's reading of these passages is, in my opinion, open to serious doubt.

The conclusion which I wish to draw from the evidence to which I have here alluded is that Locke, throughout his career, was an atomist, and that he accepted both the truth and the scientific usefulness (or, at least, the scientific promise) of the corpuscular, or new experimental, philosophy. Neither the early fragment on anatomy (1668) nor the late passage in the essay on education (1693) throw doubt on the fact that Locke was, at these times, an atomist. Between these two dates, and even subsequently, there are clear indications of his acceptance of an atomistic view of matter: they are to be found in *Draft B* (1671), in *Draft C* (1685), and in approximately equal measure in all editions of the *Essay*, from the first, in 1690, to the fourth (1700), which was the last which was published during Locke's lifetime.[32] And, to repeat my prima facie argument against Professor Yost, it would seem strange that Locke never explicitly challenged the atomistic assumptions of his contemporaries among the scientists—nor did he ever qualify his praise of their "mighty designs"—if in point of fact he doubted the utility

[32] It is to be noted that I find no particular developmental transition in Locke's thought so far as his general epistemological views are concerned, but simply a working out of them in greater detail.

Strangely enough, Thompson, whose *Study of Locke's Theory of Ideas*, attempted to trace a development in Locke's thought, fails to explain why, if there were this development, Locke left so much of his earlier thought in the later editions. It is also to be noted that the discovery of *Draft A* undercuts a good deal of Thompson's thesis. Even before the discovery of *Draft A*, however, the early materials contained in King, and used by Fraser, might have forewarned Thompson that much of Locke's supposedly later thought (which Thompson holds arose out of his concern with nominal vs. real essences) was in fact present from his earliest concern with the problem of language.

of the atomistic view of physical objects. Therefore, I shall take it as fixed throughout the remainder of this discussion that Locke can be interpreted as an atomist.

II

One reason why Locke's atomism has received so little attention may perhaps be found in the fact that atomism seems to be incompatible with a number of other doctrines which are usually regarded as being most characteristic of his thought. In the first place, atomism seems to be incompatible with the view that all knowledge has its source in sensation and reflection, for the "insensible" (i. e., imperceptible) parts of matter cannot, by definition, be presented to us in sensory experience, and knowledge of such particles cannot, of course, be gained through acts of reflection (i. e., through introspection). It would therefore seem that Locke could not be true to his own theory of knowledge and also accept atomism as a correct theory of the nature of bodies. In the second place, in his well-known distinction between primary and secondary qualities Locke states that "the ideas of primary qualities of bodies are resemblances of them, and their patterns do really exist in the bodies themselves;"[33] yet no atomist can consistently hold that the specific qualities which we perceive when we look at or when we touch material objects are identical with the qualities which these objects, when considered as congeries of atoms, actually do possess. For example, the continuous contour which characterizes the perceived shape of an object such as a table cannot be considered by an atomist to be a wholly adequate representation of that object's true shape. Now, since it is indisputable that Locke *did* draw a distinction between primary and secondary qualities, and since he also insisted that our ideas of primary qualities "resemble" these qualities in a way in which our ideas of secondary qualities do not, it is easy to assume that his atomism should not be taken seriously. In the third place, one would not expect a genuine atomist to have made the

[33] *Essay*, Bk. II, Ch. VIII, Sec. 15 (I, 173).

statements which Locke did make when he analyzed our notion of "substance," nor to have insisted, as he did insist, on the unknowability of the real essences of material objects. In short, in each of these areas of his thought, Locke's atomism would appear to be incapable of reconciliation with his fundamental epistemological convictions. What I shall now attempt to show is that this is not the case. I shall, however, start with the second apparent conflict, that concerning his doctrine of primary qualities, leaving until later the more general question of how, if at all, Locke could reconcile his acceptance of atomism with his views regarding the origin of all knowledge.

Turning, then, to the passage in which Locke states that "the ideas of primary qualities of bodies are resemblances of them, and their patterns do really exist in the bodies themselves" we must note that this passage is usually interpreted to mean that, for Locke, "the ideas of the primary [qualities] are exact representations of these qualities." [34] As we have already noted, the passage so interpreted is surely inconsistent with an acceptance of atomism. However, it is also to be noted that Locke's famous sentence is by no means unambiguous, since the notion of "resembling" and the notion of "being a pattern" leave considerable latitude in the relationship which could obtain between our ideas of primary qualities and those qualities themselves. To be sure, in some passages Locke speaks as if the quality and the idea were actually identical, but he cannot have meant this, for when he speaks cautiously he always distinguishes between an idea which is in us and a quality which is in a body. (In fact, he explicitly warns us [35] that even when he speaks incautiously we are not to understand him as meaning that ideas are in the things themselves.) Thus the question arises as to how close a resemblance there is between an idea of a primary quality and that quality as it exists in the object which possesses it.

Unfortunately, Locke is never really explicit with respect to this

[34] Aaron, *John Locke*, p. 116.
A similar interpretation is adopted by Prichard in *Knowledge and Perception*, p. 115, and by Broad in *Scientific Thought*, p. 282.
[35] *Essay*, Bk. II, Ch. VIII, Sec. 8 (I, 169).

point. I have noted some thirteen apposite cases in which he makes use of the concept of "resemblance," and in many of them he couples this term with the notions of "similitude" or "likeness," or with the term "images"; [36] yet most of these cases permit of alternative interpretations. The most usual interpretation, as we have noted, is that Locke believed that the idea of a primary quality is a direct image of that quality, resembling it as perfectly as, say, a plaster cast might resemble the statue from which it was cast. [37] Yet I do not believe that Locke actually held this doctrine. The clearest indication that he did not do so comes from the famous passage in which he says: "Had we senses acute enough to discern the minute particles of bodies, and the real constitution on which their sensible qualities depend, I doubt not but they would produce quite different ideas in us." [38] To be sure, Locke, then goes on to list illustrations of how, under these conditions, our ideas of the color of objects would be changed; however, the shape too would be changed, as he recognizes in the following section, when he says:

If that most instructive of our senses, seeing, were in any man a thousand or a hundred thousand times more acute than it is by the best microscope, things several millions of times less than the smallest object of his sight now would then be visible to his naked eyes, and so he would come nearer to the discovery of the texture and motion of the minute parts of corporeal things. [39]

---

[36] For these cases, cf. *ibid.*, Secs. 7, 13, 15, 16, 18, 22, and 25; and Bk. II, Ch. XXX, Sec. 2. (Since some of these sections contain several relevant uses of these terms I have mentioned a total of thirteen cases, but I place no special emphasis on this number.)
In addition, as we shall see, Locke's use of the term "pattern" is of importance, and in the famous sentence which I have quoted he links the notions of "resemblance" and of "patterns." For his use of "patterns" and "archetypes," cf. also Bk. II, Ch. XXX, Sec. 2; Bk. II, Ch. XXXI, Sec. 3; Bk. II, Ch. XXXII, Secs. 16, 18, and 26; Bk. III, Ch. IV, Sec. 17; Bk. III, Ch. V, Secs. 3 and 12; Bk. III, Ch. VI, Sec. 5; Bk. IV, Ch. IV, Secs. 5, 8, 11, and 12.

[37] I do not suggest that the resemblance would be as perfect as that between two statues cast from the same mould, since I assume that ideas are to be interpreted as "mental entities," and the "stuff" of which they are composed would therefore be different from that of the qualities of bodies.

[38] *Essay*, Bk. II, Ch. XXIII, Sec. 11 (I, 401).

[39] *Ibid.*, Sec. 12 (I, 403). In the same section (p. 402) Locke had already

In the light of this pasage, and especially in the light of Locke's often repeated insistence that God created our organs of sense (as well as all of our other faculties) for the ordinary concerns of our life, and not that we might achieve perfect knowledge,[40] it is difficult to accept the conventional view that he believed that our ideas of the primary qualities of macroscopic objects exactly resemble these qualities as they exist in the objects themselves.

Furthermore, if this were Locke's doctrine with respect to what he means by the primary qualities of objects, it would be extremely odd to find him holding that the powers of objects to affect other objects,—as fire affects the consistency or color of wax or of clay— are due to the primary qualities of these objects: the power of fire must be held to depend upon "the bulk, texture, and motion of its insensible parts," [41] not upon anything which exactly resembles the qualities which we directly perceive it as possessing.

Finally, we may note that Locke is willing to suggest an account of our visual perception of "the extension, figure, number, and motion of bodies of an observable bigness," and this account invokes the action of "singly imperceptible bodies" (i. e., particles) which come from the objects to our eyes and convey a motion to our brains.[42] Such an account of the origins of our ideas of the so-called primary qualities of macroscopic objects surely demands that we relinquish the view that our ideas of the qualities of these objects are replicas of the qualities as they exist in the bodies themselves: that which exists independently of us, and causes our ideas of the primary qualities of an object, is not itself capable of being perceived. In fact, throughout these important sections of Chapter VIII it is clear that Locke's real criterion of what constitutes a primary quality in

---

said: "Were our senses altered, and made quicker and acuter, the appearance and outward scheme of things would have quite another face to us." This statement cannot, in its context, be taken as applying to the so-called secondary qualities only.

[40] *Ibid.* (p. 402); cf. also Bk. II, Ch. XXX, Sec. 2, and Bk. II, Ch. XXXI, Sec. 2, as well as " The Epistle to the Reader."

[41] *Ibid.*, Bk. II, Ch. VIII, Sec. 10 (I, 171).

[42] *Ibid.*, Bk. II, Ch. VIII, Sec. 12 (I, 172). Cf. "An Examination of P. Malebranche's Opinion of Seeing All Things in God," Sec. 9–15 (*Works*, IX, 215–19).

an object is not to be ascertained by asking which of our ideas
resemble the qualities of the object itself; it is to be ascertained by
asking which of the qualities of bodies *produce* ideas in us. The latter
are the primary qualities of bodies, as Locke continually insists.[43]
And that is why in these sections he can also call these qualities the
original qualities of bodies.

This was precisely the doctrine held by Boyle, who also spoke of
"the primary qualities" as "the original qualities." (Boyle also
referred to them as "primitive," as "absolute," and as "the catholick
affections of matter," all of these terms being synonymous for him.)[44]
Now, Boyle contrasted these qualities with "sensible qualities," and
the latter included *all* of the qualities perceived by sense, that is,
they included shape and size no less than color and warmth. In fact,
Boyle said: "We must not look upon every distinct body that works
upon our sense as a bare lump of matter of that bigness and outward
shape that it appears of; many of them having their parts curiously
contrived, and most of them perhaps in motion too." [45] This position,
it seems to me, is inescapable for any atomist, and was in fact also
Locke's position.[46]

Looked at in this light we can, I believe, make far more sense of
Locke's doctrine than is usually done.[47] The primary qualities are

[43] *Draft C* is especially clear on this point. (Cf. Aaron, *John Locke*, p. 62 f.)
[44] Cf. *Works*, III, 15, 16, 22, 24, 35, 292; also, I, 308; II, 37; IV, 73, 75,
78. For his use of the term "secondary quality," cf. I, 309, and III, 24.
[45] *Works*, III, 24.
[46] It is perhaps relevant to cite a passage from the *Essay*, Bk. II, Ch. XXXI,
Sec. 6, although the special context of that passage is such that I would not
wish to place too much weight on it. Locke says:

> The particular parcel of matter which makes the ring I have on my
> finger is forwardly by most men supposed to have a real essence, whereby it
> is gold; and from whence those qualities flow which I find in it, viz. its
> peculiar color, weight, hardness, fusibility, fixedness, and change of colour
> upon a slight touch of mercury, &c. This essence, from which all these
> properties flow, when I inquire into it and search after it, I plainly perceive
> I cannot discover . . . . For I have an idea of figure, size, and situation of
> solid parts in general, though I have none of the particular figure, size, or
> putting together of parts, whereby the qualities above mentioned are pro-
> duced. (I, 508–9. My italics added.)

[47] An exception to this stricture is to be found in Reginald Jackson's article,
"Locke's Distinction between Primary and Secondary Qualities." I agree

those which produce all of our ideas of objects; they produce our ideas of the so-called secondary qualities as well as our ideas of the primary qualities. But having said this we may now ask what these primary qualities are like: do any of our ideas resemble them? And it is here that Locke answers that in all matter, constituting its primary qualities, there are certain general characteristics which we also find in certain aspects of our sensory experience, e. g., bulk, figure, number, and motion. According to Locke, there are two grounds on which these characteristics can be argued to be "utterly inseparable from body, in what state soever it be." [48] First, because "sense constantly finds [them] in every particle of matter which has bulk enough to be perceived." Second, because "the mind finds [them] inseparable from every particle of matter" even though it may be too small to be perceived. In other words, it is on the basis of generalizations, and not through immediate sensory experience, that we hold such qualities to be inseparable from matter. Thus, Locke believes (though with what right I shall not here inquire) that we may validly say that those material objects which act upon us to produce our ideas, do actually possess the primary character-istics of bulk, figure, number, and motion; however, he does not identify these characteristics as they exist in such objects with the specific ideas of bulk, or figure, or number, or motion, which their action upon us causes us to have.

Now, it may be asked why, if my interpretation be correct and

---

with Jackson when he says: "Locke means by ' primary qualities of bodies ' simply *qualities of bodies* . . . he calls them ' primary ' to distinguish them not from other qualities as a kind of qualities, but from what are on his view only wrongly thought to be qualities" (pp. 57–8). Jackson then adds that for Locke these qualities are imperceptible, and with this too I agree. However, I am inclined not to share Jackson's view on a number of other points. Nonetheless, my disagreements with Jackson's interpretations are relatively minor compared with our fundamental agreement, viz. that Berkeley and subsequent philosophers have misinterpreted Locke's doctrine of the primary qualities. Cf. also Jackson's second article on Locke, "Locke's Version of the Representative Doctrine of Perception."

[48] *Essay*, Bk. II, Ch. VIII, Sec. 9 (I, 169). This phrase is somewhat differ-ent from that contained in the first edition, but the change does not seem relevant to my point. The subsequent argument based on sense and reason is in the first edition as well as in later editions.

the real criterion of what constitutes a primary quality should be identified with that which *produces* our ideas, and not with that which is *like* certain of our ideas, Locke nonetheless so frequently speaks in terms of "resemblances." The answer is clear: he was contrasting the ontological status of these original or primary qualities with the status of the so-called secondary qualities. The latter are not qualities at all, for they exist only as ideas in us.[49] Furthermore, they do not resemble their causes: we would never know what the bulk, figure, motion, (etc.) of the insensible particles were like merely by examining our ideas of blueness, sweetness, warmth, (etc.).[50] On the

[49] Cf. *ibid.*, Ch. XXXI, Sec. 2 (I, 503), where Locke says:

Since were there no fit organs to receive the impressions fire makes on the sight and touch, nor a mind joined to those organs to receive the ideas of light and heat by those impressions from the fire or sun, there would yet be no more light or heat in the world than there would be pain if there were no sensible creature to feel it, though the sun would continue just as it is now, and Mount Aetna flame higher than ever it did. Solidity and extension, and the termination of it, figure, with motion and rest, whereof we have the ideas, would be really in the world as they are, whether there were any sensible being to perceive them or no: and therefore we have reason to look on those as the real modifications of matter, and such as are the exciting causes of all our various sensations from bodies.

In this passage it should be clear that while Locke would say that our ideas of solidity and extension "resemble" qualities which exist in bodies independently of our perception of them, his emphasis is not at all placed on the resemblance between a specific idea of, say, the shape of the sun and the shape which it possesses independently of our perception: rather, he is concerned with the problem of what types of qualities exist in nature and what types of qualities are mind-dependent. That this is also Locke's view in Chapter VIII of Book II will be argued in some detail below.

At this point, however, we may also adduce as evidence the fact that this was precisely Boyle's view as well. In a passage cited by Fraser, but without a reference, Boyle said: "If there were no sensitive beings in existence, bodies that are now the objects of our senses would be *dispositively* endowed with colors, tastes, &c; but *actually* only with those more catholic affections, as figure, motion, texture, &c., which are called primary" (*apud* note 4 to page 170 of volume I, of the Fraser edition of Locke's *Essay*). Fraser's own treatment of Locke's view in his earlier book is ambiguous, but seems strongly to suggest that he took the traditional view of what Locke meant by the "resemblance" of our ideas of the primary qualities to the qualities themselves (cf. Fraser, *Locke*, pp. 199–201).

[50] This is clearly stated by Locke in a passage in *Draft C* (cf. Aaron, *John Locke*, p. 63).

other hand, by examining our ideas of the shape and bulk (i. e., solidity) and motion of a snowball, we *can* know what shape, bulk, and motion mean when they are predicated of atoms: the "patterns" of perceived shape, bulk, and motion do exist in the objects. In this sense, perceived objects *resemble* their causes with respect to the so-called primary qualities and not with respect to the so-called secondary qualities. The fact that the shape which we perceive may not be identical with the shape which is a quality of the object does not obliterate this distinction between primary and secondary qualities, for not only is there a lack of identity between the perceived color (or sound, or taste) and any quality in the object, there is nothing which resembles color or sound or taste in the object itself. There is, as Locke has it, no *pattern* for these ideas within the object.

This interpretation of Locke's doctrine may perhaps be considered to involve so radical a departure from what has usually been taken for granted that further textual evidence in its favor should be adduced. Let us therefore start with a passage which may seem to be especially damaging, the opening of Section 18 of Book II, Chapter VIII. There Locke states:

> A piece of manna of a sensible bulk is able to produce in us the idea of a round or square figure; and by being removed from one place to another, the idea of motion. This idea of motion represents it as it really is in manna moving; a circle or square are the same, whether in idea or existence, in the mind or in the manna. And this, both motion and figure, are really in the manna, whether we take notice of them or no.

Taken in isolation, this passage might seem to demand an acceptance of the traditional interpretation of Locke's doctrine of primary qualities, that is, that our ideas of these qualities exactly resemble the qualities as they exist in the objects themselves. However, if we examine what comes immediately before this statement, and what comes after it, we can see that Locke is concerned with a different problem: he is attempting to distinguish between those types of *sensibilia* which represent, and those which do not represent, the types of qualities which are present in independently existing objects. In other words, I wish to contend that in this passage Locke is not dealing with the problem of the extent to which specific perceptual experiences may be said to be veridical, but with the question of the extent to which certain types of ideas represent the types of qualities which exist in these objects independently of our perception of them. That this is Locke's concern in the above statement can be made plausible in a series of steps.

First, such an interpretation is suggested by the sentences which comprise Section 17, the immediately preceding section of this chapter. These sentences read:

> The particular bulk, number, figure, and motion of the parts of fire or snow are really in them,—whether any one's senses perceive them or no: and therefore they may be called *real* qualities, because they really exist in those bodies. But light, heat, whiteness, or coldness, are no more really in them than sickness or pain is in manna. Take away the sensation of them; let not the eyes see light or colours, nor the ears hear sounds; let not the palate taste, nor the nose smell, and all colours, tastes, odours, and sounds, *as they are particular ideas*, vanish and cease, and are reduced to their causes, i. e., bulk, figure, and motion of parts.

I do not see that these sentences can be taken to mean that Locke is here insisting that our ideas of the primary qualities of objects faithfully and in all cases reproduce the specific qualities of these objects as they exist independently of us. The contrast which he is drawing between the primary and the so-called secondary qualities does not rest upon the fact that our ideas of the latter fail to be accurate and consistent in their delineation of objects; he is not arguing, for example, that what we see at one time as red, or taste as sweet, may at some other time appear purple, or taste bitter to us. Rather, his concern is to deny that redness or sweetness, purple or bitter, ever exist in nature independently of our perception of them. And the point of his argument is that the primary qualities of objects (qualities such as bulk, number, figure, and motion) do so exist.[51] Thus, the difference between the two classes of ideas is a difference with respect to their relevance for a description of the characteristics of

---

[51] Cf. note 49 above.

those objects which exist independently of our perceptions, and which, by their actions on our sense organs, are responsible for all of the ideas of sensation which we possess. That this really is Locke's concern in this passage, and that he is not concerned with whether our specific ideas of the macroscopic properties of objects are veridical, can also be seen by noting that in this passage he is speaking of the "bulk, number, figure, and motion *of the parts* of fire or snow" (the italics are mine): these parts actually possess such qualities "whether any one's senses perceive them or no." Therefore, I do not believe it plausible to hold that in Section 17 Locke is to be interpreted as putting forward a doctrine concerning the relation between the perceived shape of an object and the real shape of that object.

Nonetheless, as we have noted, whatever may be the case with respect to Section 17, the opening sentences of Section 18 deal with "a piece of manna of sensible bulk" and seem to be directly concerned with the question of perceived shape. However, in order to interpret these sentences correctly we must also look at what follows immediately upon them. This examination constitutes the second step in my attempt to show what Locke actually wishes to hold in the passage in question.

Immediately after stating that "motion and figure are really in the manna, whether we take notice of them or no," Locke goes on to state that manna (the laxative, and not the heavenly nourishment) "by the bulk, figure, texture, and motion of its parts, has a power to produce the sensations of sickness, and sometimes of acute pains or gripings in us." The point which he wishes to make in this passage is that whiteness and sweetness are not to be taken as residing in the manna any more than are the sickness and pain which it is capable of causing in us. In other words, Locke is reverting to precisely the same point which he made in Section 17 concerning the difference in ontological status between the primary qualities and the so-called secondary qualities. Considering the context of the intervening sentences, it would seem odd to interpret them in such a way as to lend support to the traditional view of Locke's doctrine: one would have to assume that in the midst of a discussion of what characteristics are possessed by physical objects independently of the reactions of our organisms to them, and what characteristics are not to be regarded as existing independently of those reactions, Locke suddenly introduced three sentences concerning the epistemological question of whether our ideas of shape and motion exactly resemble what is to be found in these objects. I do not say that Locke (because of a confusion, or for some other reason) might not have done precisely this. However, before assuming that he did do so, it would be well to see if some other interpretation of the critical passage might not be at least as plausible as that interpretation which upholds the traditional view of Locke's theory.

It will be recalled that in the sentences here at issue Locke says: "A piece of manna of sensible bulk is able to produce in us the idea of a round or square figure; and by being removed from one place to another, the idea of motion." This sentence can, without strain, surely be taken to signify that from the observation of bodies which are sufficiently large to be sensed, we derive our ideas of roundness, of squareness, of figure in general, and of motion. Thus far, then, nothing has been said concerning the resemblance of the idea of the particular shape of this sensed object to its actual, or inherent, shape, nor of any correspondence between its perceived motion and its actual motion. However, Locke then appears to raise the latter question, for he says: "this idea of motion represents it as it really is in the manna moving." However, the syntax of this phrase is difficult: to what does the word "it" refer? I can only understand Locke to mean what might be paraphrased in saying: "This idea of motion, which we have derived from observing a piece of manna of a sensible bulk being moved from one place to another, represents a characteristic which truly qualifies a piece of manna when it is moved." In short, the passage may, I submit, be interpreted as stating that our idea of motion, which we have derived from experience, does represent a characteristic which exists in nature independently of our experience. And when this poorly constructed phrase is taken in this sense, the following phrase takes on similar meaning, for Locke says: "a circle or square are the same, whether in idea or existence, in the mind or in the manna." And Locke immediately continues in the same vein: "And this [i. e., these characteristics], both motion and figure, are really in the manna, whether we take notice of them or no." In none of these statements do I therefore find any reason to assume that Locke should be interpreted as saying that the specific shape which we perceive exactly resembles the true shape of the object. On the contrary, Locke's statements, for all their peculiar syntax, seem to me more naturally interpreted if he is held to be discussing the ontological status possessed by that which we term figure or motion, and that he is only contending that such ideas represent features of an independent physical world, whereas sickness and pain, and whiteness and sweetness, do not.

I have labored over the interpretation of this small segment of Locke's chapter on the primary and secondary qualities in order to show that what might at first glance seem to disprove my interpretation of his doctrine is in fact wholly consistent with it. There are, however, other passages in which Locke holds that our ideas of the primary qualities of macroscopic objects *do* faithfully depict characteristics of these objects; his distinction between the primary and the secondary qualities of objects is not therefore to be taken as *merely* applying to the insensible parts of bodies. We must now see why this is the case.

Let us assume with Locke that the insensible parts of bodies all necessarily possess qualities of extension, solidity, figure, and mobility, but that they do not, in themselves, possess any qualities corresponding to color, or taste, or sound. If, now, a number of these minute particles of matter come together, may we not speak of that group of particles as itself possessing extension, or solidity, or figure, or mobility? It would surely seem so, for some such groups occupy a larger region of space than do others, and some will resist penetration or separation more than do others; therefore, it is wholly natural to speak of these groups of particles as extended and as being solid in varying degrees, etc. In short, the qualities which Locke attributed to each particle of matter are also attributable to groups of such particles. He was therefore able to say that our ideas of the extension, solidity, figure, and mobility of macroscopic objects resemble what exists in these objects independently of us, and thus that our ideas of the primary qualities of macroscopic objects really resemble qualities possessed by those objects as they exist independently of us.

Nonetheless, I wish to insist that this concern with the perceived size, shape, solidity, or mobility of sensed objects is not the basic aspect of Locke's doctrine of the primary qualities: it is really only an addendum to his main point, which concerns the fact that these qualities exist in the insensible particles of all material objects. This is perhaps most clearly brought out in Section 22 when Locke states:

> I have in what just goes before been engaged in physical inquiries a little further than perhaps I intended. But, it being necessary to make the nature of sensation a little understood; and to make the difference between the *qualities* in bodies, and the *ideas* produced by them in the mind, to be distinctly conceived, without which it were impossible to discourse intelligibly of them; —I hope I shall be pardoned this little excursion into natural philosophy; it being necessary in our present inquiry to distinguish the *primary* and *real* qualities of bodies, which are always in them (viz., solidity, extention, figure, number, and motion, or rest, and are sometimes perceived by us, viz., when the bodies they are in are big enough singly to be discerned), from those *secondary* or *imputed* qualities, which are but the powers of several combinations of those primary ones, when they operate without being directly discerned.

The upshot of our argument, which is well summarized by the above summary statement made by Locke himself, is that the basis on which Locke established his theory of the primary qualities was his atomism; it was not his aim to attempt to establish the nature of physical objects by examining the sensible ideas which we had of them. Thus, instead of viewing Locke's doctrine of the primary and secondary qualities as a doctrine which rests on an analysis of differences among our ideas, his doctrine is to be understood as a theory of physical entities, and of the manner in which our ideas are caused.[52] To this extent the Berkeleian criticism of Locke's distinc-

---

[52] I have noted only one passage in which Locke's distinction between the primary and the so-called secondary qualities might be supposed to rest not on causal analysis, but on features of ideas considered merely as elements within our experience. This is to be found in the *Essay*, Bk. II, Ch. VIII,

tion between primary and secondary qualities is wholly beside the point, for it rests on an assumption which Locke did not share—that all distinctions concerning the nature of objects must be based upon, and verified by, distinctions discernible within the immediate contents of consciousness.[53]

It may seem mistaken to hold that Locke did not intend to distinguish between primary and secondary qualities in terms of a distinction among the immediate data of consciousness, for his discussion of the issue is to be found at the end of his treatment of simple ideas, and the chapter in which it occurs is entitled "Some Further Considerations Concerning Our Simple Ideas of Sensation." Were my view to be accepted, one might think that Locke should not have placed his discussion of the qualities of bodies at this point, but should only have introduced such a discussion in connection with our complex ideas of the nature of substances. And this, in fact, is true: the chapter in question does actually belong with his discussion

---

Sec. 21, where he contrasts the consistency of the testimony of our senses with respect to primary qualities with the contradictions among our ideas of the secondary qualities. He says:

> If we imagine *warmth* as it is in our hands, to be nothing but a certain sort and degree of motion in the minute particles of our nerves or animal spirits, we may understand how it is possible that the same water may, at the same time, produce the sensations of heat in one hand and cold in the other; which yet *figure* never does, that never producing the idea of square by one hand which has produced the idea of a globe by another.

However, the context of this passage is an examination of how our ideas of the secondary qualities depend upon the effects on our organisms of the "texture" of the insensible parts of objects, and the contention that we are not deceived by tactile impressions of shape plays no significant part in the discussion—it is simply mentioned and dropped. In other passages, for example in his discussions of the "reality" of our simple ideas (Bk. II, Ch. XXX, Sec. 2) and in his discussion of the "adequacy" of our simple ideas (Bk. II, Ch. XXXI, Sec. 2), the distinction between our ideas of the primary and secondary qualities is not said to rest upon differences to be found within these ideas considered as ideas, but upon knowledge of their causes.

[53] Berkeley would, of course, attack Locke's reliance upon physical theory, seeking to prove that it too must also rest on data which are confined to the immediate contents of consciousness. With that argument I am not here concerned: I only wish to point out that in so far as the Berkeleian position rests on identifying Locke's distinction between primary and secondary qualities with a distinction between two types of ideas of macroscopic objects, it is wholly unfair to Locke's argument as well as to his aim.

of the nature of individual substances, and not primarily with his discussion of simple ideas.

Oddly enough, this point is almost always overlooked: among the better-known commentaries, only Gibson departed from Locke's own order of exposition and discussed the question of the primary qualities in connection with his discussion of Locke's view of substances. Yet in *Draft B*, to which other more recent commentators have had access, the two topics are discussed together, as I would claim that they should have been. Now, actually it is not difficult to see precisely how Locke, subsequently to the writing of *Draft B*, came to separate them by fourteen intervening chapters.[54] Chapter VIII opens with a discussion of the fact that all of our simple ideas are "positive" ideas; i. e., it opens with Locke insisting that an idea which depends upon "a privative cause" is no less a positive and simple idea than one which depends upon some active power in the object: cold is no less a simple, positive idea than heat, nor black than white, etc. Thus, the six opening sections of Chapter VIII fit naturally into the end of Locke's discussion of simple ideas. And if we ask why it was necessary to introduce this discussion, the answer clearly lies in Locke's recognition of the fact that when we are discussing ideas *qua* ideas we should not draw distinctions between them on the basis of their causes. As he says in the comparable passage in *Draft B*, "it being one thing to know the idea of black or white, and quite another to examine what kind of particles it must be, and how ranged in their superficies, to make it appear black."[55] Thus it was not unreasonable (if such a warning were necessary, as Locke clearly believed that it was)[56] that he should have added this

---

[54] *Draft C* also separates the discussion of primary and secondary qualities from the discussion of our complex ideas of substances. Aaron's account of the relevant passages of *Draft C* (Aaron, *John Locke*, pp. 61–63) may suggest that there was an additional reason for the introduction of Chapter VIII where it is, viz., that Chapter VII had a more extended discussion of our idea of power, and included a mention of the primary attributes of matter and spirit. However this may be, the reasons which will here be adduced are of themselves sufficient to explain the order of these chapters.

[55] *Draft B*, p. 118.

[56] In the *Essay* (Bk. II, Ch. VIII, Sec. 3) the warning is explicitly directed against those "philosophers" (i. e., natural scientists) who study the theory of colors. After discussing this point, Locke immediately turns to warn others

discussion at the end of his general treatment of simple ideas. On the other hand, it is impossible to make this point without presupposing, as Locke obviously does, that some ideas depend upon active powers, while others are caused in us by the relative absence of these powers.[57] This presupposition involves distinguishing between *ideas* (as being in our minds) and *qualities* (as being in bodies), and this distinction immediately leads to a discussion of the primary, or original, qualities of bodies and the so-called secondary qualities. Thus, having introduced the discussion of the positive nature of all of our ideas into the discussion of simple ideas, Locke was necessarily led into a discussion of the nature of material substances. But why, then, we may ask, did he not immediately go on in the *Essay* (and in *Draft C*), as he had in *Draft B*, to discuss the nature of substances? The answer to this lies in the fact that to talk of the distinction between a quality in an object and an idea in our minds is to raise questions concerning the roles of sensation and judgment in perception, and thus Locke was led into his chapter concerning perception, and from there to his other psychological chapters on memory, etc. The latter, after all, concern what the mind does with the simple ideas, and they have a bearing on the theory of how our complex ideas are formed; they are therefore by no means out of place. Nonetheless, it is necessary to insist that we cannot ultimately separate what Locke has to say concerning the primary qualities of material objects from what he has to say concerning our complex ideas of substances, and it is in my opinion unfortunate that most commentators have done precisely this. I wish then to turn to the question of how Locke's chapter entitled "Of Our Complex Ideas of Substances" it to be interpreted.

---

not to equate their ideas with the qualities of objects (*ibid.*, Sec. 7). Thus, like anyone upholding a representative theory of perception, he must argue against both the naïve realism of common sense and the tendency of some scientists to substitute a physical cause for a perceived quality.

[57] In Section 4 of this chapter (I, 167), Locke even suggests a psychophysical explanation of how "privative causes" affect us.

III

It would be presumptuous to hold that the reason why most commentators have failed to link Locke's discussion of primary and secondary qualities with his doctrine of substance is only to be found in the fact that the chapter on substance is separated by fourteen other chapters from that dealing with qualities and powers. Not only would commentators such as Aaron not have been misled by this fact, but the chapter on substance itself reverts to the distinction.[58] The failure to link these chapters rests, rather, on the persistence of the view that the most essential component in Locke's analysis of our complex ideas of material substances is the notion of an unknown substrate which underlies and supports the sensible qualities of these objects. This interpretation of what is most essential in Locke's doctrine seems to me fundamentally misleading.

In order to introduce an alternative interpretation of the twenty-third chapter of Book II of the *Essay*, let me first call attention to the fact that both in it and elsewhere Locke distinguishes between what he terms "substance in general" and what he designates as "particular sorts of substances."[59] Whenever he uses the singular form, "substance," or when he speaks of "pure substance in general," he is referring to an unknown and unknowable substratum; whenever he uses the plural form he is speaking not of our conception of a substratum, but of individual things, or of types of individual things. So far as I am aware, there is no passage in which Locke confuses these two distinct notions;[60] however, many of his interpreters have unfortunately failed to follow his example.

Now, it is to be noted that the title of Locke's chapter uses the plural form, "Of Our Complex Ideas of Substances." It should also be noted that only in the first five sections of this chapter is Locke primarily concerned with the notion of a substratum. If we examine the immediately subsequent sections (Sections 6 through 14) we

[58] Cf. *Essay*, Bk. II, Ch. XXIII, Sec. 8 (I, 399).
[59] E. g., Bk. II, Ch. XXIII, Sec. 3 (I, 392).
[60] Locke himself insists very strongly on this distinction in his first letter to Stillingfleet. Cf. *Works*, IV, 17.

find that Locke's attention is confined to the question of how we come to know the characteristics of particular substances of varying types; for example, when he alludes to our knowledge of the properties of gold, iron, horse, man, vitriol, bread, the sun, water, diamonds, and loadstones, the notion of a substratum plays no part. In fact, this is Locke's basic contention: that the notion of a substratum or "pure substance in general" gives us no knowledge of the properties of particular substances. For example, in *Draft C* he said:

> We have no idea of the *substance* of body or any other thing, but it lies wholly in the dark, because when we talk of or think on those things which we call natural substances, as man, horse, stone, the idea we have of either of them is but the complication or collection of those particular simple ideas of sensible qualities which we use to find united in the thing called *Horse* or *Stone*.[61]

And in the *Essay* itself it is clear that he is concerned to attack those who placed an undue emphasis on the notion of substance in general, instead of analyzing the particular characteristics of particular types of substances. In this he was surely not only attacking Descartes but also those who stood in the Aristotelian-Scholastic tradition; unlike both of these schools of thought, he rejected the centrality of the category of substance, which for him did not represent a clear and determined idea. One can see his impatience with current uses of the term "substance" when he says:

> It helps not our ignorance to feign knowledge where we have none, by making a noise with sounds, without clear and distinct significations. Names made at pleasure neither alter the nature of things, nor make us understand them, but as they are signs of and stand for determined ideas. And I desire those who lay so much stress on the sound of these two syllables, *substance*, to consider whether applying it, as they do, to the infinite, incomprehensible God, to finite spirits, and to body, it be in the same sense;

---

[61] *Apud* Aaron, *John Locke*, p. 60. This passage is closely paralleled by the summary statement of his doctrine in the *Essay*, Bk. II, Ch. XXIII, Sec. 4, which is quoted in full in note 69, below. However, the emphasis in the two is slightly different: *Draft C* better illustrates the negative side of Locke's doctrine with which I am here concerned.

and whether it stands for the same idea, when each of those three so different beings are called substances.[62]

And in the same section, and in the same context, he speaks of "the promiscuous use of so dubious a term." Even clearer, is the following ironical passage:

> Whatever a learned man may do here, an intelligent American, who inquired into the nature of things, would scarce take it for a satisfactory account, if, desiring to learn our architecture, he should be told that a pillar is a thing supported by a basis, and a basis something that supported a pillar. Would he not think himself mocked, instead of taught, with such an account as this? And a stranger to them would be very liberally instructed in the nature of books, and the things they contained, if he should be told that all learned books consisted of paper and letters, and that letters were things inhering in paper, and paper a thing that held forth letters: a notable way of having clear ideas of letters and paper. But were the Latin words, *inhaerentia* and *substantio*, put into the plain English ones that answer them, and were called *sticking on* and *under-propping*, they would better discover to us the very great clearness there is in the doctrine of substance and accidents, and show of what use they are in deciding of questions in philosophy.[63]

In the light of such passages one can scarcely think that it was Locke's primary concern in his chapter on substances to prove that our conception of particular substances involves the notion of an unknowable substratum.

What has served to focus undue attention on the problem of the substratum is the supposed inconsistency between Locke's theory that all ideas have their sources in sensation and reflection and his admission that in thinking of particular substances we think of their qualities as inhering in an unexperienced and unexperienceable substratum. Stillingfleet used this apparent inconsistency to attack Locke's "way of ideas"; Berkeley, on the other hand, later used it to attack Locke's realism. Yet Locke did not alter his views after

---

[62] *Essay*, Bk. II, Ch. XIII, Sec. 18.
[63] *Ibid.*, Sec. 20 (I, 230 f.).

Stillingfleet's attack; in fact, he apparently made only one slight change of wording in his discussion of the substratum in the fourth edition of the *Essay*, after he had had his exchanges with Stillingfleet.[64] That he did not find it necessary to alter his doctrine rests on the fact that his position was not really inconsistent: he was always perfectly ready to admit that our notion of a substratum (or of substance in general) does not come to us by either sensation or reflection, and he could admit this because we do not possess a "particular distinct positive idea" regarding it.[65] As he said in his first letter to Stillingfleet:

I never said that the general idea of substance comes in by sensation and reflection; or that it is a simple idea of sensation or reflection, though it be ultimately founded in them: for it is a complex idea, made up of the general idea of something, or being, with the relation of a support to accidents. For general ideas come not into the mind by sensation or reflection, but are the creatures or inventions of the understanding.[66]

Throughout his discussions of the problem Locke emphasized the limitations of this general idea: he characterized it as "obscure and relative," not clear and distinct;[67] further, he characterized it as being a *supposition*, not something of which we have a sensible idea. Therefore, Locke's use of the notion of a substrate is not incompatible with his doctrine of the origin of all simple ideas in sensation and reflection.

To be sure, there are a number of other points at which Locke's discussion of the substratum does lead him into difficulties and confusions. Perhaps the most salient of these is the fact that throughout his discussion of the substrate he fails to take into account his own distinction between ideas as they exist in the mind and the qualities and powers which are to be attributed to objects. For example, in

[64] Cf. Bk. I, Ch. III, Sec. 19 (I, 108, n. 1).
[65] *Ibid.*, p. 108.
[66] *Works*, IV, 19.
[67] E. g., *Essay*, Bk. II, Ch. XXIII Sec. 3. Also, in his first letter to Stillingfleet, Locke characterized the idea of a substratum in saying: "I have a very confused, loose, and undetermined idea of it, signified by the name substance" (*Works*, IV, 29).

the first and third sections of Chapter XXIII his analysis of our belief in a substrate is couched in terms of "ideas"; yet, in the intervening section it is couched in terms of "qualities." Associated with his failure in this respect there is also a confusion between our belief in the substratum as being that which explains why certain sets of *ideas* go constantly together, and a belief in the substratum as that in which *qualities* inhere.[68] As the context of Section 1 of this chapter makes clear, his introduction of the substratum as an explanation of why these sets of ideas do go together is related to Locke's realism: his insistence that something must stand behind what is given in experience, causing our sensations. In Section 2 this reason is repeated in terms of the scholastic doctrine of accidents, but to it there is added the notion that every *quality* must inhere in a substance which serves as the ground of explanation for it. Had Locke clearly avoided a confusion between our ideas and the qualities of objects, these passages would have had to be radically revised.[69]

[68] To speak (as Locke does) of the substratum of a material object as being that in which *ideas* subsist is clearly a confusion: only qualities and powers could be spoken of as subsisting in material objects. It is also a confusion to speak as if the relation of "inhering in" were equivalent to the relation of "resulting from," as Locke does when he says in Section 1: "we accustom ourselves to suppose some *substratum* wherein [these simple ideas] do subsist, and from which they do result, which therefore we call *substance*."

[69] This double confusion is evident in Locke's summary statement which makes up Section 4 of this chapter:

> Hence, when we talk or think of any particular sort of corporeal substances, as horse, stone, &c., though the idea we have of either of them be but the complication or collection of those several simple ideas of sensible qualities, which we used to find united in the thing called horse or stone; yet *because we cannot conceive how they should subsist alone, nor one in another*, we suppose them existing in and supported by some common subject; which support we denote by the name substance, though it be certain we have no clear or distinct idea of that thing we suppose a support. (I, 395)

In the phrase italicized by Locke, "they" presumably refers to "several simple ideas of sensible qualities," but once one distinguishes between ideas and qualities it is nonsense to hold that these ideas exist in, and are supported by, anything in the horse or the stone. Therefore, when Locke says that "we cannot conceive how they should subsist alone" he is only saying that we cannot conceive how these ideas should arise in us as they do were they not caused by qualities which actually inhere in the particular substances of which

Nonetheless, after admitting that there are these confusions in Locke's exposition of his doctrine, let us see if we cannot make better sense of it if we consistently follow his injunction to distinguish between ideas as they are in the mind and qualities and powers as these exist in things.[70] In doing so, let us close the gap separating Locke's discussion of qualities and powers from his discussion of our complex ideas of substances, and consider his doctrine of substance in the light of his discussion of the primary and secondary qualities of objects. Above all, in offering our interpretation let us utilize the sharp distinction which Locke himself drew between the notion of substance in general and our complex ideas of particular substances.

Our interpretation must start from the fact that Locke is perfectly clear on one point: in our ordinary experience our complex idea of any particular object involves the possession of a whole group of determinate ideas, some of which are simple ideas, and some of which (while being equally specific and determinate) are in fact complex relational ideas which can (according to Locke) be treated "for brevity's sake" as if they were simple ideas. As examples of the former we may cite the color and consistency of an object; examples of the latter are our ideas of the changes in color or consistency which a particular type of object undergoes under varying conditions.[71] To this whole set of determinate and specific ideas (whether simple or not simple) we add the supposition of a substratum. Unlike these determinate ideas, the notion of the substratum is indeterminate and has no place in direct experience; yet it also is included in our overall conception of what constitutes a particular substance. Furthermore, in addition to these two types of components we must take cognizance of a third set of factors which Locke introduces into his discussion of our ideas of particular sub-

---

we form our ideas. It is in this way that his original confusion of idea and quality, plus his realism, led him to confuse inherence with causation. Such a confusion may well have been facilitated by a failure on Locke's part to see the full difference between his own views on explanation and earlier assumptions that the causal relation could often (or always) be interpreted in terms of what followed from the nature of a substance. (Cf. below, p. 52, and note 104; also, p. 59 f.)

[70] Cf. *Essay*, Bk. II, Ch. VIII, Sec. 7–8.

[71] Cf. *ibid.*, Ch. XXIII, Sec. 7 (I, 397–98).

stances: namely, those qualities and powers which exist in these substances independently of the sensible effects through which we become cognizant of them.[72] As we have seen, such qualities and powers are not to be equated with our sensible ideas, according to Locke. Thus, contrary to the usual interpretation of Locke's doctrine of substances, in which (following his own error) the distinction between ideas and qualities is not consistently maintained, an analysis of our various beliefs about particular substances will involve three different types of components, not two. We shall have to take into account all of the specific sensible ideas which we have concerning such substances; wè shall also have to include the supposition of a substratum; and, third, we shall, in addition, have to take into account the specific qualities and powers which reside in the substances themselves, in so far as we can determine what these qualities and powers actually are.

In order to interpret Locke's doctrine we must now see how these three discriminable sets of components are related to one another. First we may note that in Chapter XXIII, no less than in Chapter VIII, Locke always assumes that our ideas of objects are caused by the action of the primary qualities on us, and that these primary qualities are the qualities of the particles of which material objects are composed.[73] In the second place, as we have already seen, Locke assumes that one of the reasons why we are driven to suppose the existence of a substrate is that we cannot avoid believing that there is something which causes our ideas of sensation.[74] Taken in conjunction with the previous proposition this suggests that our notion of a substratum is connected with the notion of the inner, atomic constitution of objects, since both are regarded by Locke as being causally related to our ideas of the sensible qualities of things. To this rather surprising linkage of the notions of an unknown and

---

[72] Cf. *ibid.*, Sec. 9, where Locke reintroduces his classification of the qualities and powers of objects, repeating what he had said in Ch. VIII.

[73] Cf. *ibid.*, Secs. 9 to 12.

[74] It must be noted that Locke's own analysis of how we come by our ideas of cause and effect (*ibid.*, Ch. XXVI, Sec. 1) and of power (Ch. XXI, Sec. 1) do not really permit him to hold this view. Nonetheless, regardless of the inconsistency, there can be no doubt that he does so.

unknowable substratum and of the atomic constitution of material objects we shall later return.[75] Here, however, we must in the third place note that, in spite of this linkage, the substrate and the primary and original qualities of objects cannot be equated with one another in any simple, straightforward fashion. One significant difference between them is that no analysis of the qualities and powers of an object in any way clarifies the indeterminate general idea of a substrate; unlike the inner constitution of things, the substrate remains merely "a something," and this locution "signifies no more, when so used, either by children or men, but that they know not what."[76] Thus, the second and third of our remarks seem to stand in conflict with one another, the second suggesting a close affiliation between the substratum of an object and the internal constitution upon which its powers depend, whereas the third involves a fundamental opposition between them.

I suggest that this conflict can be at least partially resolved if we make the hypothesis that in his analysis of the role of the substrate in our complex ideas of substances Locke was confining his attention to what we take objects to be like on the basis of our ordinary daily experience with them; in other words, that in discussing the notion of substance in general he was not considering objects in the light of what could be discovered about them by the methods of the new experimental philosophy. On this interpretation our supposition of a substratum enters into our complex idea of a particular substance, but it does so as an idea only: it is not to be identified with the

[75] In Section 3 of Chapter XXIII there is one passage which can be taken as evidence—though not conclusive evidence—that Locke actually was linking these notions in his own mind. In speaking of how we take note of the fact that certain ideas regularly accompany one another, Locke says that we therefore suppose them "to flow from the particular internal constitution, or unknown essence of that substance." If we may take "the unknown essence" to be equivalent to the substratum in this passage—as I believe the context permits us to do—we have textual evidence in the *Essay* itself for connecting the general notion of a substrate and the atomic constitution of material objects.

As another example of a case in which Locke is apparently linking the unknown substratum and the ultimate physical constitution of bodies, Section 2 of *Draft A* may be cited: in it he equates the substrate with "substance or mater."

[76] *Essay*, Bk. II, Ch. XXIII, Sec. 2.

properties of the object itself. Its place in our ordinary conception
of an object is, so to speak, that of a surrogate for what in the object
is material and exists independently of us—i. e., that which is not
merely an idea or group of ideas. Put in different words, our conception of a substratum is an indeterminate and general notion
standing for something in the object which makes that object a
self-subsisting thing, that is, a thing which (in Cartesian language)
needs nothing else in order to exist.[77] Thus, so far as our ordinary
experience is concerned, this vague notion of a substrate corresponds
to the atomist's view of the role of atoms as the original and unchanging matter on which all sensible appearances depend. However, our indeterminate notion of a material substrate stands in need
of correction by inferences based on the observation of the powers
of objects: it is the atomic constitutions of objects, not "pure substance in general," which cause the ideas of them which we actually
have, and which also cause the effects, whether perceived or unperceived, which objects have upon one another. Thus, no appeal
to the notion of substance in general—no use of this category—
will solve the problem of what objects are like. Yet Locke believes
that we inescapably form this indeterminate general notion, since
he holds—as his later discussion of sensitive knowledge clearly
shows [78]—that we are inescapably realists in our ordinary experience
and cannot doubt that in perceiving or acting we are being affected
by material objects which are independent of us.

To this interpretation of Locke's doctrine of the substrate it might
be objected that ordinary men do not ordinarily think in terms of
some substrate underlying specific sensible qualities; in other words,
that such a notion is not part of our conception of the everyday
world of material objects. I would agree that it is not, and that in
this sense Locke was undoubtedly wrong in his description of our
experience. Nonetheless, when one considers the extent to which
his predecessors had emphasized the notion of substance as being

[77] Locke speaks of the substratum as "that unknown common subject,
which inheres not in anything else." And he speaks of it as the cause of the
union of our ideas "as makes the whole subsist of itself." (Both quotations
are from the *Essay*, Bk. II, Ch. XXIII, Sec. 6.)
[78] *Ibid.*, Bk. IV, Ch. XI, Secs. 2 and 3.

that which underlies all qualities and events, it is not surprising that Locke was misled by a philosophical theory when he came to describe our everyday experience of the world. Furthermore, as we have noted, the sections of the chapter which are concerned with the notion of the substrate have a polemical and negative cast: in them Locke is attempting to argue that through the use of this indeterminate and general notion we can get no concrete knowledge of the nature of material objects. It is only through the determinate ideas of the relations of objects to one another, and through noting how our simple sensible ideas of them change under changing conditions, that we can obtain reliable knowledge of what exists independently of us. Thus, the burden of this chapter is that we must appeal to experience, and not to the notion of substance in general, to ascertain the nature of objects. The experience to which we must appeal is, in the first instance, our ordinary observation in daily life. However, Locke points out that continuous with this everyday knowledge there are the sorts of observations made by smiths and by jewelers, and such observations, as we know from Boyle, were themselves regarded as being continuous with the experimental philosophy, in which more refined and systematic inquiry serves to establish the various powers of specific types of objects. In support of the fact that this is Locke's view, it is to be noted that in his discussion of our knowledge of particular substances he is led to talk of what we would see if we had microscopical eyes, of what the blood is really like, of how we experimentally establish the nature of gold, etc. It is in the light of such knowledge of the specific natures of specific types of bodies that he ridicules relying upon a notion of substance in general as constituting any part of an explanation of what occurs in nature. To be sure, he admits—and in fact insists—that in our perceptual experience we always find ourselves using this notion of something underlying whatever surface qualities we observe an object to have; however, ordinary perceptual experience, while useful in all of the concerns of life, does not for Locke reveal the nature of material objects as they are in themselves.[79]

[79] Cf. *ibid.*, Bk. II, Ch. XXIII, Secs. 12 and 13.
Whether Locke believed that we would also think in terms of a substrate

As a final justification for this interpretation of Locke's doctrine of substance I shall now show how it fits with what he says concerning the real and the nominal essences of substances.

It is in Book III of the *Essay* that Locke draws his distinction between the real and the nominal essences of substances. One of the passages in which he does so reads as follows:

> Essence may be taken to be the very being of anything, whereby it is what it is. And thus the real internal, but generally (in substances) unknown constitution of things, whereon their discoverable qualities depend, may be called their essence . . . .
>
> [However,] the learning and disputes of the schools having been much busied about *genus* and *species*, the word *essence* has almost lost its primary signification: and, instead of the real constitution of things, has been almost wholly applied to the artificial constitution of *genus* and *species*. It is true, there is ordinarily supposed a real constitution of the sorts of things; and it is past doubt there must be some real constitution on which any collection of simple ideas co-existing must depend. But, it being evident that things are ranked under names into sorts or species,

if we had "microscopical eyes" is not clear. The one passage which might be taken to bear on this problem might suggest that he believed that we would. However, the passage is ambiguous since one cannot be sure that in it Locke is speaking of the characteristics of the atomic particles, or whether he has in mind the solidity and extension of molar matter. The passage reads as follows: "If any one should be asked, what is the subject wherein colour or weight inheres, he would have nothing to say, but the solid extended parts; and if he were demanded, *what is it that solidity and extension adhere* in, he would not be in a much better case than the Indian . . . [etc.]" (Bk. II, Ch. XXIII, Sec. 2; italics mine).

This passage may seem to lend support to an interpretation of Locke's doctrine of substance which has been advanced by John Yolton, but which seems to me fundamentally mistaken. Yolton claims that "a physical object for Locke was defined as being composed of three elements: secondary qualities or powers, primary qualities, and substance or substratum" ("Locke's Unpublished Marginal Replies to John Sargent," p. 555). I think it a mistake to regard the notion of a substratum as being connected with the *actual qualities* of an object, rather than as being an indeterminate notion connected with our sensible *ideas* of such qualities. The fact that Yolton takes the former alternative forces him to say that "the real essence of any physical object is hidden away in the unknowable but necessary substratum." As we shall see immediately below, this would seem to contravene Locke's own words as to what constitutes the real essence of a material object.

only as they agree to certain abstract ideas, to which we have annexed those names, the essence of each *genus*, or sort, comes to nothing but that abstract idea which the general . . . name stands for . . . .

These two sorts of essences, I suppose, may not unfitly be termed, the one the *real*, the other *nominal essence*.[80]

Locke illustrates the difference between these two meanings of the term "essence" in the following way:

The nominal essence of gold is that complex idea the word gold stands for, let it be, for instance, a body yellow, of a certain weight, malleable, fusible, and fixed. But the real essence is the constitution of the insensible parts of that body on which those qualities and all the other properties of gold depend.[81]

This distinction between the nominal and the real essences of things is identical with the distinction which I have just drawn between Locke's account of our ordinary notions of particular substances and what he believes that an analysis of objects in terms of their internal parts would disclose. He himself draws such a contrast between two ways of looking at things when, in Book III of the *Essay*, he says:

For, though perhaps voluntary motion, with sense and reason, joined to a body of a certain shape, be the complex idea to which I and others annex the name *man*, and so be the nominal essence of the species so called: yet nobody will say that complex idea is the real essence and source of all those operations which are to be found in any individual of that sort. The foundation of all those qualities which are the ingredients of our complex idea, is something quite different: and had we such knowledge of that constitution of man . . . we should have a quite other idea of his species, be it what it will: and our idea of any individual man would be as different from what it is now, as is his who knows all the springs and wheels and other contrivances within the famous clock at

---

[80] Bk. III, Ch. III, Sec. 15 (II, 26–27). One finds the same distinction in Boyle, and many of Locke's illustrations resemble his; however, Boyle does not use the same terminology. (For relevant passages, cf. *Works*, III, 18–19, 27.)
[81] *Essay*, Bk. III, Ch. VI, Sec. 2.

Strasburg, from that which a gazing countryman has of it, who barely sees the motion of the hand, and hears the clock strike, and observes only some of the outward appearances.[82]

Thus, the nominal essences of things are not to be considered as being like their real essences. In other words, the ordinary way in which we form our conceptions of substances through putting together a series of simple ideas which regularly accompany one another (and by adding the supposition of a substrate in which they inhere) does not provide us with ideas corresponding to the actual mode of existence of material things.[83]

If this be doubted, consider Locke's attack on the Aristotelian-Scholastic doctrine that species exist in nature.[84] He regards our distinction between species as artificial, that is, as being made by men, and not found in nature independently of man. As he says:

> Our distinct species are *nothing but distinct complex ideas, with distinct names annexed to them.* It is true that every substance that exists has its peculiar constitution, whereon depend those sensible qualities and powers we observe in it; but the ranking of things into species (which is nothing but sorting them under several titles) is done by us according to the ideas that *we* have of them: which, though sufficient to distinguish them by names, so that we may be able to discourse of them when we have them not present before us; yet if we suppose it to be done by their real internal constitutions, and that things existing are distinguished by nature into species, by real essences, according as we

---

[82] *Ibid.*, Sec. 3. As we shall see, Boyle uses the illustration of the Strasburg clock in a similar connection (cf. below, p. 90 f.), and Locke himself again refers to it in Section 9 of this same chapter.

[83] Perhaps the clearest indication of Locke's desire to avoid assigning to our complex ideas of substances any status in the independent physical world is to be found when he points out the difference between using the word "gold" as a generic name signifying "the complex idea which I or any one else calls gold" and using it to refer not to an idea but a thing, "a particular piece of matter, v. g. the last guinea coined" (cf. *ibid.*, Sec. 19).

[84] Locke's attack on the independent reality of species was in many places couched in terms of biological facts, and must have been of at least indirect influence in undermining the biological theory of the invariance of species. (For some of his discussions of this topic, cf. *ibid.*, Ch. III, Secs. 13 and 17, and Ch. VI, Secs. 12, 16, 17, 22, 23, 26, 27, 29, and 34.)

distinguish them into species by names, we shall be liable to great mistakes.[85]

Locke of course found those who stood in the Aristotelian-Scholastic tradition guilty of this type of mistake: it was this identification of species with real essences that he rightly regarded as underlying their attempts to explain natural events in terms of substantial forms.[86] In fact, almost the whole of this chapter, "Of the Names of Substances," may be considered as an attack on that tradition. However, it is no less an attack on the adequacy of our ordinary, common-sense view that sensory experience teaches us the true differences which exist between different types of material objects. As Locke repeatedly points out, our sorting of things into distinct species, classifying them and naming them according to linguistic traditions and the needs of life, does not give us a correct conception of their natures.[87] However, the limitations of our knowledge of material objects does not rest on this alone. According to Locke there are at least three other reasons why the sensible properties on the basis of which we distinguish among material objects of various sorts do not serve as adequate indicators of the true natures, or real essences, of these objects. In the first place, as we have noted, our sensible ideas do not accurately depict the actual physical constitu-

---

[85] *Ibid.*, Sec. 13.

[86] Cf. *ibid.*, Sec. 10.

[87] Cf. *ibid.*, Secs. 29 and 30, part of whose argument may be given in the following truncated form:

Where we find the colour of gold, we are apt to imagine all the other qualities comprehended in our complex idea to be there also . . . . But though this serves well enough for gross and confused conceptions . . . it requires much time, pains, and skill, strict inquiry and long examination to find out what, and how many, those simple ideas are, which are constantly and inseparably united in nature, and are always to be found together in the same subject. Most men, wanting either time, inclination, or industry enough for this, even to some tolerable degree, content themselves with some few obvious and outward appearances of things, thereby readily to distinguish and sort them for the common affairs of life (II, 80-81).

Similarly, he says: "It is their own collections of sensible qualities that men make the essences of *their* several sorts of substances; and that their real internal structures are not considered by the greatest part of men in the sorting of them" (*ibid.*, Sec. 24).

tion of these bodies. In the second place, as we here find Locke insisting, and as he frequently insists elsewhere, our experience of the particular characteristics of bodies is limited, and never exhaustive. Therefore, since all of the characteristics of an object depend upon its real essence, the limitations of our experience will preclude us from saying that we have discovered what constitutes the real nature of these objects. In the third place, as Locke also suggests, there may well be powers in objects by virtue of which they affect other objects, and are affected by them, which we never suspect "because they never appear in sensible effects." [88] And, in fact, Locke believes that a good part of the nature of any object depends upon its covert relations with other objects, as organisms depend upon their environment; yet in framing a complex idea of these objects we are prone to overlook these relations. As he says:

> We are wont to consider the substances we meet with, each of them, as an entire thing by itself, having all its qualities itself, and independent of other things; overlooking for the most part, the operations of those invisible fluids they are encompassed with, and upon whose motions and operations depend the greatest part of those qualities which are taken notice of in them, and are made by us the inherent marks of distinction whereby we know and denominate them . . . . The qualities observed in a loadstone must needs have their source far beyond the confines of that body; and the ravage made often on several sorts of animals by invisible causes . . . evidently show that the concurrence and operations of several bodies, with which they are seldom thought to have anything to do, is absolutely necessary to make them be what they appear to us, and to preserve those qualities by which we know and distinguish them. We are then quite out of the way, when we think that things contain *within themselves* the qualities that appear to us in them.[89]

Thus once again we see that Locke has drawn a distinction between our complex ideas of substances and the actual nature of material substances as they exist independently of us. The former are com-

[88] *Ibid.*, Bk. II, Ch. XXIII, Sec. 9 (I, 400).
[89] *Ibid.*, Bk. IV, Ch. VI, Sec. 11 (II, 260–61).

posed of a congeries of simple ideas of sensation, plus the supposition of a substratum in which those ideas inhere, and to which we
affix a name; the latter are objects which possess qualities distinct
from the sensible ideas which they cause in us, and bear no necessary relationship to the classificatory schemes under which we are
apt to arrange them.

If this be true, how then—one may ask—can Locke claim to know
that there are substances distinct from us, or know anything concerning what their natures may be like?

IV

Before answering these questions, and thus answering the fundamental question of this essay—in what measure Locke may be said
to have been consistent in his realism—it will be well to offer a more
general interpretation of Locke's thought than the present study
has yet contained.

In this connection we should first take cognizance of Locke's
purpose in writing the *Essay*. What he sought to ascertain was "the
origin, certainty, and extent of human knowledge." This search was
not, however, motivated simply by a theoretical interest in the problem of knowledge as such. In the first place, as is evident from his
friend James Tyrell's remark,[90] and as is also surely clear from
internal evidence in the drafts of the *Essay*, Locke was interested
in the problem of knowledge for the light it could throw on the
possibility of settling disputes which concerned moral and religious
questions.[91] This early concern seems to be echoed in a significant
passage near the end of the *Essay*:

> Since our faculties are not fitted to penetrate into the internal
> fabric and real essences of bodies; but yet plainly discover to us
> the being of a God, and the knowledge of ourselves, enough to

[90] Cf. Aaron and Gibb, *An Early Draft of Locke's Essay*, p. xii.
[91] For example, in *Draft A*, immediately following a discussion of substances,
there is a very brief paragraph (#3) on relations in general, but this is then
followed by an extended paragraph (#4) on moral relations.

lead us into a full and clear discovery of our duty and great concernment; it will become us, as rational creatures, to employ those faculties we have about what they are most adapted to, and follow the direction of nature, where it seems to point us out the way. For it is rational to conclude, that our proper employment lies in those inquiries, and in that sort of knowledge which is most suited to our natural capacities, and carries in it our greatest interest, i. e., the condition of our eternal estate. Hence I think I may conclude, that *morality* is *the proper science and business of mankind in general.*[92]

However, in addition to this moral concern, there was another powerful motive which led Locke to undertake his prolonged analysis of human knowledge: he wished to correct the false pretensions of system-builders and dogmatists. This motivation is clear as early as Locke's fragment, *De Arte Medica*, dated 1668.[93] In it system-building, and the idle terminological disputes and dogmatic assertions connected with system-building, are characterized by Locke as the chief obstacles to the improvement of useful knowledge; and system-building in its turn is regarded as springing from man's pride of intellect. As Locke said:

. . . True knowledge grew first in the world by experience and rational observations; but proud man, not content with the knowledge he was capable of, and which was useful to him, would needs penetrate into the hidden causes of things, lay down principles, and establish maxims to himself about the operations of nature, and then vainly expect that nature, or in truth God, should proceed according to those laws which *his* maxims had prescribed to him; whereas his narrow and weak faculties could reach no further than the observation and memory of some few facts produced by visible external causes, but in a way utterly

[92] Bk. IV, Ch. XII, Sec. 11.
[93] The fragment is given in Bourne, *The Life of John Locke*, I, 222 ff., but the relevant passage is also given in Fraser's edition of the *Essay*, I, xxiv f. The date of this fragment is significant in that it shows Locke's basic reason for a concern with the problem of knowledge some three years before the discussion out of which the *Essay* sprang. However neither it nor the other medical fragments seem to me to justify Romanell's view that the *Essay* originated in problems of medical methodology.

beyond the reach of his apprehension;—it being perhaps no absurdity to think that this great and curious fabric of the world, the workmanship of the Almighty, cannot be perfectly comprehended by any understanding but His that made it. Man, still affecting something of the Deity, laboured by his imagination to supply what his observation and experience failed him in; and when he could not discover (by experience) the principles, causes and methods of nature's workmanship, he would needs fashion all these out of his own thought, and make a world to himself, framed and governed by his own intelligence. This vanity spread itself into many useful parts of natural philosophy; and by how much the more it seemed subtle, sublime, and learned, by so much the more it proved pernicious and hurtful, by hindering the growth of practical knowledge . . . .

This distrust of human pretensions to infallible knowledge, this interest in what is of practical concern to man, and this contempt for purely theoretical systems, are perhaps most clearly evident in an extended entry which Locke made in his Notebook in 1677 while he was in Montpellier, working on the *Essay*.[94] They are, of course, also reflected, though in softer focus, in the "Epistle to the Reader," which stands as Locke's introduction to the *Essay*. But one might wonder why, if these were Locke's real concerns, the first draft of the *Essay* should have started with a discussion of how our notions of particular substances are formed by a compounding of simple ideas of sense, to which there is added the vague idea of a substrate. Why should the problem of what constitutes our knowledge of a material object such as the sun, or of what comprises our knowledge of gold, be the starting point of Locke's discussion?

In the first place, we may note that Locke is here concerned to show how we get our knowledge of concrete entities with which we are concerned in the ordinary affairs of life. He is not concerned with the abstract nature of matter, nor is he concerned to discuss cosmology: we gain our knowledge of the physical world through experience of specific objects. In the second place, he is accounting for what we know of the nature of particular substances in terms of

[94] This entry is reprinted in Aaron and Gibb, *An Early Draft of Locke's Essay*, pp. 83–90. It is also given in King, *The Life of John Locke*, I, 161–71.

experienced sensible qualities which they can be observed to have, not in terms of substantial forms. In the third place, in his account of our conception of particular substances, and of types of substances, he introduces the effect of names on our notions of particular substances, showing how these names, though useful, may mislead us. And, finally, he shows us that the more complete our knowledge of the powers of things—e. g., of the ductility of gold—the more we may be said to know of them. In all these respects what he has to say reminds one immediately and unmistakeably of Boyle's concerns when Boyle was arguing for the new corpuscular or experimental philosophy.

That there should be this connection between Locke's starting point in the first draft of the *Essay* and Boyle's philosophy is not strange. And that Locke had in mind defending this philosophy— and the method of work of the virtuosi of the Royal Society—is not only to be expected from his remark on Sydenham, Huyghens, Boyle, and Newton in the "Epistle to the Reader," but can be documented by his Notebook entry of 1677 to which I have just referred. In that entry he does insist that the mind of man "findes it self lost in the vast extent of space, and the least particle of matter puzzles it with an inconceivable divisibility"; he also admits that perhaps man's mind cannot know "the essence of things, their first originall, their secret way of workeing and the whole extent of corporeall beings," nor "the nature of the sun and stars . . . and 1000 other speculations in nature." However, here as elsewhere, Locke also insists that "this state of our mindes however remote from that perfection whereof we our selves have an Idea, ought not however to discourage our endeavours in the search of truth or make us thinke we are incapeable of knowing any thing because we cannot fully understand all things." And in fact what Locke finds us chiefly incapable of understanding are "the more generall and forain parts of nature," not what we have experience of. This passage is worth quoting at length:

> . . . what need have we to complaine of our ignorance in the more generall and forain parts of nature when all our bisinesse lies at hand why should we bemoane our want of knowledg in the par-

ticular apartments of the universe when our portion lies only here in this litle spot of earth, where we and all our concernments are shut up. Why should we thinke our selves hardly dealt with that we are not furnishd with compasse nor plummet to saile and fathom that restlesse and innavigable Ocean of the Universall matter motion and space since if there be shoars to bound our voiage and travaile, there are at least noe commoditys to be brought from thence serviceable to our uses nor that will better our condition, and we need not be displeasd that we have not knowledg enough to discover whether we have any neighbours or noe in those large bulks of matter we see floating in that abysse, and of what kinde they are since we can never have any communication with them nor enterteine a commerce that might turne to our advantage . . . .[95]

The knowledge which can be turned to our advantage is twofold: natural and moral knowledge. And of natural knowledge Locke singles out for attention such knowledge as can provide us with the means of life and can improve our material lot.

> Here then is a large feild for knowledg proper for the use and advantage of men in this world viz To finde out new inventions of dispatch to shorten or ease our labours, or applying sagaciously togeather severall agents and patients to procure new and beneficiall productions whereby our stock of riches (i. e., things usefull for the conveniencys of our life) may be increased or better preservd. And for such discoverys as these the minde of man is well fitted.[96]

If with this passage in mind one recalls that Boyle, when writing to Marcombes, spoke of the Royal Society as "our new philosophical college that values no knowledge but as it hath a tendency to use." [97] And if one also recalls that Boyle and his colleagues considered the

---

[95] Journals appended to *Draft A*, in Aaron and Gibb, *An Early Draft of Locke's Essay*, p. 86. (Also to be found in King, *The Life of John Locke*, I, 165.)

[96] Journals appended to *Draft A*, in Aaron and Gibb, *An Early Draft of Locke's Essay*, p. 85. (Also in King, *The Life of John Locke*, I, 163.)

[97] Letter dated Oct. 22, 1646, *apud, Record of the Royal Society of London*, p. 3.

new science as the way to achieve useful inventions, then one can see that there is no conflict between this practical aim and Locke's satisfaction that even though men must remain ignorant of so much, we do have "abilitys to improve our knowledge in experimentall naturall philosophy." [98] What this new philosophy can achieve is to give us a more correct knowledge of *particular substances* (*not* of "pure substance in general"), and this knowledge, like the philosophers' stone, could transform and control nature for our advantage.

Supposing, then, that we interpret Locke as being, in his fundamental philosophic motivation, a follower of this Baconian tradition, a tradition which included Boyle among its most eminent examples: what can we then say concerning Locke's views regarding our knowledge of substances?

In the first place we must note a distinction which Locke draws between "the proper science and business of mankind in general" and "the lot and talent of particular men." In the passage from his chapter entitled "Of the Improvement of our Knowledge" which I have already cited, [99] he had insisted that "our faculties are not fitted to penetrate into the internal fabric and real essences of bodies," and from this he had argued that the proper employment of these faculties lay in the field of morality. However, in stating this conclusion he drew an explicit contrast between what is true of "mankind in general" and what is true of certain individuals, saying: "I think I may conclude that *morality* is *the proper science and business of mankind in general* . . . as several arts, conversant about several parts of nature, are the lot and private talent of particular men." The results of the cultivation of these arts, as he then goes on to show by using the illustration of the discovery of iron, are of inestimable value for the generality of mankind. Having said this, he adds, "I would not, therefore, be thought to disesteem or dissuade the study of *nature*." Thus, we must read much of Locke's discussion concerning our inability to know the real essences of bodies as being directed against any claim that our senses and our common experience

---

[98] Journals appended to *Draft A*, in Aaron and Gibb, *An Early Draft of Locke's Essay*, p. 88.
[99] *Essay*, Bk. IV, Ch. XII, Sec. 11; cited above, pp. 46 f.

furnish us with such knowledge—that such knowledge is, in other words, open to the generality of mankind.[100] If this distinction is legitimate, as I take it to be, it would go far to substantiate my reading of Locke's discussion of substances, for according to that interpretation (it will be recalled) there was at least an implicit contrast between the origins of our ideas of bodies as these were derived from immediate sensible ideas, and the real essences of bodies with which experimental philosophy sought to deal.

To be sure, Locke consistently denies that *anyone*—whether plain man or scientist—can know the real essence of any individual material object. However, to interpret his insistence on this point, we must once again note the manner in which he uses the term "knowledge."[101] For him "knowledge," or "science," was to be distinguished from "opinion" and "probability": that which was not certain could not be characterized as science or as knowledge. Apart from the special case of "sensitive knowledge,"[102] knowledge for Locke consisted in the immediate intuitive or the mediate demonstrative perception of the agreement or disagreement among our ideas. However, Locke saw no way in which any such knowledge concerning material substances could be established.[103] For Descartes,

---

[100] As we have previously noted (pp. 13 f., above), Locke's *Essay* is not a treatise on the nature, scope, and limits of *scientific method*; it is directed primarily to showing the origin, certainty, and extent, of what we, today, are apt to call "the plain man's" knowledge.

If I am not mistaken, the opening sentences of Section 14 of Book IV, Chapter XII, should also be read in the light of this distinction between everyday knowledge and the results of scientific inquiry. That sentence reads: "But whether natural philosophy be capable of certainty or no, the ways to enlarge our knowledge, as far as we are capable . . . . (etc.)." On my interpretation, Locke is here drawing a contrast between "natural philosophy" and "*our* knowledge."

[101] Cf. above, p. 12.

[102] "Sensitive knowledge," for Locke, is not knowledge of the particular qualities which external objects possess, but is only the belief (which he takes to be justified) *that* there is an external world causing our simple ideas of sensation, and that these ideas, therefore, represent the action on us of something which is independent of us.

[103] At times, by inadvertence, he did use the term "knowledge" in a broad sense which included probability as well as "knowledge" proper. For example, in his chapter on the improvement of our knowledge (*Essay*, Bk. IV, Ch. XII), he used the term to cover the results of our inquiries into the nature of material

and for the Aristotelians, at least some of the properties of objects were explicable in terms of the nature of the substances whose properties they were. This was true of the non-accidental properties for the Aristotelians. It was also characteristic of Descartes' ideal of knowledge in which effects were to be explained through their causes, that is, properties through the substances which served as their grounds.[104] However, Boyle reversed this order, and insisted that knowledge proceeds from effects to causes;[105] knowledge for him was to be observational and empirical, not rational. And, so far as particular material substances were concerned, Locke wholly agreed with Boyle's method. It was because he agreed with this method, and because he none the less adopted the stricter Cartesian definition of "knowledge," that Locke refused to characterize our information concerning material bodies as knowledge.[106]

substances as well as to apply to the demonstrative sciences of morals and mathematics.

[104] Cf. "Rules for the Direction of the Mind," VI (in the Haldane and Ross edition of *The Philosophical Works*, I, 16); also *Principles of Philosophy*, Pt. I, Prop. XXIV (*ibid.*, I, 229).

[105] Cf. *Works*, IV, 72-73.

It is also worthy of note that Colin Maclaurin, in his *Account of Sir Isaac Newton's Philosophical Discoveries*, takes it to be a characteristic difference between Newton and Descartes that the Cartesians "express contempt for that knowledge of causes which is derived from the contemplation of their effects, and are unwilling to condescend to any other science than that of effects from their causes" (p. 14). In this connection he cites Descartes' *Principles*, Part III, Prop. 4.

[106] For example, note the following statement which occurs in the *Essay*, Bk. II, Ch. XXXI, Sec. 6:

The complex ideas we have of substances . . . cannot be the real essence, . . . for then the properties we discover in that body would depend on that complex idea, and be deducible from it, and their necessary connexion with it be known; as all properties of a triangle depend on, and, as far as they are discoverable, are deducible from the complex idea of three lines including a space. But it is plain that in our complex ideas of substances are not contained such ideas, on which all the other qualities that are found in them do depend.

It is also to be noted that in one passage Locke relates our lack of demonstrative knowledge concerning material objects to the fact that the foundation of all of our knowledge rests on our senses, and he suggests that immaterial spirits (and of course God) could have demonstrative knowledge of the various

Finally, in order to interpret Locke's doctrine concerning the limitations of our knowledge of the real essences of material substances we must draw a distinction between knowing the general properties of such substances and knowing their individual natures. As I have attempted to show, it seems indubitable that Locke believed that all material objects were composed of atoms, and that these atoms possessed certain properties, which he termed their primary qualities. There seem to be no grounds for holding that he was ever skeptical of the warrant of this belief. On the other hand, it is no less clear that he denied that we have any means of directly discerning, or even accurately inferring, the particular sizes, shapes, number, or motions of the particles which go to make up any specific object, or even any specific type of object.[107] These two views are not, however, incompatible: we need merely draw the suggested distinction between the possibility of knowing the general properties possessed in common by all material substances, and the specific

properties and powers of material things from a knowledge of their ultimate natures—presumably from a knowledge of their insensible parts. He says:

> The whole extent of our knowledge or imagination reaches not beyond our ideas limited to our ways of perception. Though yet it be not to be doubted that spirits of a higher rank than those immersed in flesh may have as clear ideas of the radical constitution of substances as we have of a triangle, and so perceive how all their properties and operations flow from thence (Bk. III, Ch. XI, Sec. 23).

[107] For example, Locke says:

> Our faculties carry us no further towards the knowledge and distinction of substances, than a collection of *those sensible ideas which we observe in them*; which, however made with the greatest diligence and exactness we are capable of, yet is more remote from the true internal constitution from which those qualities flow, than, as I said, a countryman's idea is from the inward contrivance of that famous clock at Strasburg, whereof he only sees the outward figure and motions. There is not so contemptible a plant or animal, that does not confound the most enlarged understanding. Though the familiar use of things about us take off our wonder, yet it cures not our ignorance. When we come to examine the stones we tread on, or the iron we daily handle, we presently find we know not their make; and can give no reason of the different qualities we find in them. It is evident the internal constitution, whereon their properties depend, is unknown to us: for to go no further than the grossest and most obvious we can imagine amongst them, What is that texture of parts, that real essence, that makes lead and antimony fusible, wood and stones not? (*ibid.*, Ch. VI, Sec. 9).

properties of different, specific substances. In discussing the adequacy of our ideas of material substances, Locke himself draws this distinction, saying:

> The particular parcel of matter which makes the ring I have on my finger is forwardly by most men supposed to have a real essence, whereby it is gold; and from whence those qualities flow which I find in it, viz. its peculiar colour, weight, hardness, fusibility, fixedness, and change of colour upon a slight touch of mercury, &c. This essence, from which all these properties flow, when I inquire into it and search after it, I plainly perceive I cannot discover: the furthest I can go is, only to presume that, it being nothing but body, its real essence or internal constitution, on which these qualities depend, can be nothing but the figure, size, and connexion of its solid parts.

And in the same passage, in stating his opposition to the doctrine of substantial forms, he adds:

> I have an idea of figure, size, and situation of solid parts in general, though I have none of the particular figure, size, or putting together of parts, whereby the qualities above mentioned are produced; which qualities I find in that particular parcel of matter that is on my finger, and not in another parcel of matter, with which I cut the pen I write with.[108]

Thus, whatever may have been his skepticism regarding our ability to penetrate into the secret material constitution of *individual* things, this skepticism did not cast doubt on the acceptability of the atomistic hypothesis as a general explanation of all of the powers which we observe that particular types of bodies are capable of displaying.

Bearing these points in mind, we are now in a position to examine Locke's doctrine concerning the extent of our knowledge of physical objects, taking the term "knowledge" in that broader signification which includes all well-grounded beliefs concerning these objects. The key to Locke's analysis of such knowledge lies in what he has to say concerning "powers."

In both of his discussions of the differences between the primary

---

[108] *Ibid.*, Bk. II, Ch. XXXI, Sec. 6 (I, 507, 508).

and the secondary qualities of objects, Locke includes, in addition to these qualities, what he terms "powers." [109] While his terminology in these two passages is not identical, in both of them it is clear that the powers of an object are what we should call dispositional properties of that object; that these dispositional properties depend upon the nature of the qualities inhering in the minute parts of the objects, that is, on the object's primary qualities; and that the so-called secondary qualities are themselves to be classed among the powers of objects, being their powers to affect our sense organs in a particular way. The difference between the so-called secondary qualities of an object and its other powers is, according to Locke, the difference between something immediately affecting our sense organs, and something which does so mediately. As he says:

> We immediately by our senses perceive in fire its heat and colour; which are, if rightly considered, nothing but powers in it to produce those ideas in *us*: we also by our senses perceive the colour and brittleness of charcoal, whereby we come by the knowledge of another power in fire, which it has to change the colour and consistency of *wood*. By the former, fire immediately, by the latter, it mediately discovers to us these several powers.[110]

By classing these two sorts of powers together, Locke—as we shall now see—actually paves the way for treating our ordinary knowledge of material objects as continuous with that more refined analysis of these objects which can be attained by experimental investigators. As will be recalled, his analysis of our ordinary conceptions of material objects consists in holding that we group together whatever simple ideas of sensation regularly accompany one another, and to this we add the supposition of an unknown substratum in which these subsist. These sensible ideas now turn out, however, to be merely powers in the object causing us to have these particular sensations, rather than others. Therefore, there is no reason why Locke should not hold that when one object is observed to induce changes in some other object, this dispositional property of the first object should not be regarded as a quality of that object just as much as are its so-called secondary qualities. To be sure, in our sensory

[109] Cf. *ibid.*, Ch. VIII, Secs. 9–10, and Ch. XXIII, Secs. 9–10.
[110] *Ibid.*, Ch. XXIII, Sec. 7 (I, 398).

experience we do not ordinarily so regard it: the observed color or felt smoothness or hardness of a bar of magnetized iron seem to be "qualities" of that piece of iron in a way in which its active power to attract iron filings, or its passive power to melt at a certain temperature, do not seem to be. However, Locke classes all such properties as powers, and does so in spite of the fact that, as he recognizes, and as we have noted, the former are in fact simple ideas, whereas the latter are not.[111] This thesis on his part allows him to hold that our observations of the behavior of various types of objects under varying conditions form no less a part of an adequate complex idea of these substances than whatever sensible qualities these substances present to us immediately in sensory experience. And this of course allows him to hold that the more refined observations and investigations of various types of substances which are carried on by jewelers and by smiths, and also by chemists and other natural philosophers, are to be regarded as successive improvements upon what our own direct and untutored sensory observation reveals.

This continuity between our ordinary untutored conceptions of the properties of a particular type of substance and the almost limitless possibilities of further knowledge concerning such substances is traced out by Locke in such passages as the following:

> Whosoever first lighted on a parcel of that sort of substance we denote by the word *gold*, could not rationally take the bulk and figure he observed in that lump to depend on its real essence, or internal constitution. Therefore those never went into his idea of that species of body; but its peculiar colour, perhaps, and weight, were the first he abstracted from it, to make the complex idea of that species. Which both are but powers; the one to affect our eyes after such a manner, and to produce in us that idea we call yellow; and the other to force upwards any other body of equal bulk, they being put into a pair of equal scales, one against another. Another perhaps added to these the ideas of fusibility and fixedness, two other passive powers, in relation to the operation of fire upon it; another, its ductility and solubility in *aqua regia*, two other powers, relating to the operation of other bodies, in changing its outward figure, or separation of it into insensible parts. These, or parts of these, put together, usually make the

---

[111] Cf. above, p. 36.

complex idea in men's minds of that sort of body we call *gold*.

But no one who hath considered the properties of bodies in general, or this sort in particular, can doubt that this, called *gold*, has infinite other properties not contained in that complex idea. Some who have examined this species more accurately could, I believe, enumerate ten times as many properties in gold, all of them as inseparable from its internal constitution, as its colour or weight.[112]

However, Locke is not for this reason contemptuous of our ordinary conceptions of objects, since these are, by and large, sufficient for the needs of our daily lives. In fact, he suggests that if our senses were more acute, and enabled us directly to perceive the "secret composition and radical texture of bodies," this capacity would be inconvenient for our ordinary conduct and well-being.[113] In this connection, as always, he recommends that we be content with the position in which God has placed us. Nonetheless, as we have noted, Locke explicitly states that he does not for this reason wish to be thought to "disesteem or dissuade the study of nature," [114] and he then puts forward his view of how such knowledge should proceed:

> In the knowledge of bodies, we must be content to glean what we can from particular experiments: since we cannot, from a discovery of their real essences, grasp at a time whole sheaves, and in bundles comprehend the nature and properties of whole species together . . . . He that shall consider how little general maxims, precarious principles, and hypotheses laid down at pleasure, have promoted true knowledge, or helped to satisfy the inquiries of rational men after real improvements; how little, I say, the setting out at the end has, for many ages together, advanced men's progress, towards the knowledge of natural philosophy, will think we have reason to thank those who in this latter age have taken another course.

This is unmistakably the method of Boyle and of those others whom

[112] *Essay*, Bk. II, Ch. XXXI, Secs. 9 to 10. Cf. Bk. III, Ch. VI, Secs. 30 and 31.
[113] Cf. *ibid.*, Bk. II, Ch. XXIII, Secs. 12 and 13; also Bk. IV, Ch. XII, Sec. 11.
[114] *Ibid.*, Bk. IV, Ch. XII, Sect. 12 (II, 351 f.).

he classed as master builders, when in the "Epistle to the Reader" he defined his own task as that of clearing away some of the rubbish that lay in the path of human understanding.[115]

Now, if this praise of the new experimental philosophy is taken seriously, what can one say of Locke's earlier and apparently far more skeptical discussion of the *extent* of human knowledge? "[116] In that well-known discussion he apparently insists on certain ineradicable limitations to human knowledge: we can never be certain what qualities will coexist with what other qualities, nor can we discover how the sensible ideas which we have of objects are connected with the qualities which they possess. However, this passage is by no means incompatible with what we have been claiming to be Locke's position. The way in which his inquiry is couched in these paragraphs is in terms of whether we can ever have necessary knowledge of "the agreement or disagreement of our ideas in co-existence," and this he takes to mean whether we can see a necessary connection between various simple ideas (or ideas of powers) which go to make up our complex ideas of substances. It should occasion no surprise (nor should it be taken as in any way justifying skepticism) that, for Locke, "the simple ideas whereof our complex ideas of substances are made up are, for the most part, such as carry with them, in their own nature, no *visible necessary* connexion or inconsistency with any other simple ideas, whose coexistence with them we would inform ourselves about." [117] In short, what he is attempting to show throughout these sections is that:

> Our knowledge in all these inquiries [concerning the co-existence of certain sensible ideas] reaches very little further than our experience. Indeed some few of the primary qualities have a necessary dependence and visible connexion with one another, as figure necessarily supposes extension; receiving or communicating

---

[115] As we have had occasion to note, the chief opponents of Locke doubtless were the Scholastics. However, he (like Boyle) was also critical of the alchemists, as can be seen in a passage concerning the methods of "the philosophers by fire" which he added in the second edition to Bk. IV, Ch. III, Sec. 16. And it is possible that what he has to say about "systems," "hypotheses," and "principles" in Bk. IV, Ch. XII, Secs. 12 and 13, might have been directed against Descartes.

[116] *Ibid.*, Ch. III, Secs. 9 to 16 (II, 199–206).

[117] *Ibid.*, Sec. 10.

motion by impulse, supposes solidity. But though these, and perhaps some of our other ideas have: yet there are so few of them that have a visible connexion one with another, that we can by intuition or demonstration discover the co-existence of very few of the qualities that are to be found united in substances.[118]

In other words, we must consult experience (which means sense experience) to learn the nature of particular types of substances: we cannot discover either by intuition or by demonstration what their precise properties will be; thus we can never have certainty in our opinions concerning them. However, nowhere in this passage does Locke evince the least doubt that objects do exist in their own right, independently of us; that they possess the characteristics which atomism assigns to them; and that it is because of their atomic constitution (and doubtless also because of our natures) that they cause us to form the ideas which we do form of them.

To those who might wish to challenge the grounds on which Locke assumed that material objects existed independently of our minds, Locke himself proposed an answer in his treatment of our sensitive knowledge; and, if my interpretation of his doctrine of the substratum is correct, he also proposed a closely connected sort of answer when he claimed that, when our ideas regularly accompany one another, we cannot believe that they occur as they do without supposing some underpinning which is responsible for their concurrence. I should not wish to claim that Locke's reply to the challenges which came to be posed by Berkeley and by Hume are adequate. However, it cannot be denied that Locke did see the problem, in outline at least, and did attempt to propose a solution to it. What seems missing in his system is something else: it is the absence of any attempt to justify the acceptance of that atomism which runs throughout his discussion of human knowledge. It is my opinion that Locke did not feel obliged to justify this theory because he not unnaturally viewed it as an empirically based conclusion drawn from the experimental inquiries of his day. I say "not unnaturally," since, as we shall now see, this is also the way in which at least two of the new "master builders," Boyle and Newton, regarded the corpuscularian philosophy.

[118] *Ibid.*, Sec. 14 (II, 203); cf. Bk. IV, Ch. VI, Sec. 7, and Ch. XII, Secs. 9–10.

2

NEWTON AND BOYLE
AND THE PROBLEM
OF "TRANSDICTION"

In Order To Understand Locke's assurance regarding the existence of a world of physical objects whose natures were dependent upon their atomic constitutions, and whose inherent properties were therefore different from the sensible ideas through which we originally come to know them, we must—as I have suggested—consider the corpuscularianism of Boyle and of Newton. What will here occupy our attention is the problem of how these two natural philosophers sought to justify their beliefs concerning the ultimate constituents of the physical world, without contradicting their views regarding the role of experience in human knowledge. This, as we shall see, constitutes the problem of "transdiction."

I borrow the term "transdiction" from Professor Donald C. Williams who used it in commenting upon a paper delivered by Carl G. Hempel before the Harvard Philosophy Club in 1958. Professor Hempel had been speaking of the conditions under which one can predict or retrodict from data given at a certain time to what will happen, or to what has happened, at another time. In Professor Hempel's discussion, both the observed data and the events which were to be predicted (or retrodicted) were assumed in all cases to be either experienced or experienceable entities. Professor Williams, however, wished to use data in such a way as not only to be able to move back and forth *within* experience, but to be able to say something meaningful and true about what lay beyond the boundaries of possible experiences. This he termed "transdiction." Furthermore, he contended that as a matter of historical fact, scien-

tists in the past had generally believed in the legitimacy of transdiction. In H. H. Price's way of putting the same matter in his
discussion of causal inference in *Perception*, it was Williams' contention that physical scientists believe in "vertical" as well as in
"horizontal" causation.

If we may take Boyle and Newton as examples, there seems to
be little doubt that Professor Williams was correct in his historical
generalization. As I shall attempt to show, not only did each believe
in the legitimacy of transdiction, but each had his own method of
seeking to justify it. Before examining these historical matters, however, it will be necessary briefly to discuss the relation of transdiction
to inductive inference in general. Since this is not a question which
has been much discussed, at least in the form in which I wish
to discuss it, what I shall have to say on the problem will be both
tentative and somewhat crude.

I

If one examines recent works on logic and on the philosophy of
science it seems to be the case that the term "induction" is in-
creasingly being taken to refer to all forms of nondemonstrative
inference. For the sake of convenience I shall accept this usage
without discussion, and having accepted it we may immediately
note that transdiction is then to be classified as one form of inductive
inference. To be sure, the question will immediately arise as to
whether this form of inductive inference is justifiable. Now, this
question is not identical with what has been called "*the* problem
of induction." That problem, which is also sometimes called "Hume's
problem," is a question which corresponds to another and narrower
definition of what constitutes induction. The definition which it
presupposes equates induction with those inferences in which we
proceed from the fact that something is true of a certain number
of members of a class to the conclusion that the same thing will
be true of unknown members of that class.[1] "The problem of in-

---

[1] This was von Wright's definition of induction in the opening sentence of

duction" which arises out of this definition is whether we can justify such inferences. One of the commonest ways of raising this problem is to ask whether we have a right to assume that the future will be like the past. Another way of asking the same sort of question, without explicitly introducing the time element, is to ask whether the fact that all observed cases of, say, crows being black justifies the inference that all birds resembling crows in other respects will also resemble them in being black. Such in general, is what has been labelled "the problem of induction." Now, it may be the case (and I believe that it is the case) that whenever we are involved in drawing any nondemonstrative conclusion from empirical data, this type of question can be raised in a form relevant to that inference. And thus the so-called "problem of induction" will also be relevant to all transdictive inferences. However, this is clearly not the only problem, nor the most important problem, concerning such inferences. What I shall call "the problem of transdiction" (on the analogy of speaking as if there were only one problem of induction), is, rather, the question of how observed data can serve as grounds for inferences to objects or events which not only have not yet been observed, but which cannot in principle be observed.

The form in which this problem has most often been raised is in connection with the doctrine of epistemological realism.[2] And it would indeed seem that most, if not all, attempts to establish the existence of independent material objects as the causes of our perceptions would, at some point, involve transdiction. Yet, there are two reasons why we must not identify the problem of transdiction with the problem of establishing epistemological realism.

The Logical Problem of Induction. It has been retained in the second, revised edition of that book. In my opinion, it is more or less typical of the earlier standard characterizations of inductive inference.

[2] In *Induction and Hypothesis* (pp. 7–8), S. F. Barker recently singled out the problem of epistemological realism vs. subjectivism as one of the topics to which a study of inductive logic was relevant. In his treatment of the problem of induction in *Problems of Philosophy* (p. 93), Bertrand Russell also mentioned "the existence of matter" as one of the important questions involved in the theory of inductive inference. However, neither of these works keeps this problem steadily in mind in its actual treatments of the logic of inductive inference, and the special applications of such inference to transdictions are not treated in them.

In the first place, there is no one-to-one correlation between epistemological realism and the use of transdiction. For example, some philosophers, such as Kant, might best be characterized as phenomenalists, rather than realists, and yet it is probably not false to insist (as Hegel insisted) that at least some minimal form of transdictive inference is involved in the doctrine that we can know *that* there are entities which stand in some sort of causal relation to our sensory experience, even though we are unable to characterize their specific natures. On the other hand, there are realists who claim that we are *directly* aware of the fact that objects exist independently of our sense experience, and that we directly know the natures of these objects. Such a view, which is generally classified as "direct" or "naive" realism, presumably does not employ transdictive inference. In fact, there are no inductive inferences of any kind that this form of realism would employ in establishing realism; it would only use such inferences in an attempt to defend its position. Thus, on the one hand, some who seem to employ at least a minimal form of transdictive inference are not realists, and some who are realists would seem to deny the necessity for such inferences.

The second reason why the problem of transdiction should not be equated with the problem of epistemological realism lies in the fact that transdiction is used by scientists for specific scientific purposes, quite independently of any general epistemological concerns. As Braithwaite has pointed out, "an adequate theory of science to-day must explain how we come to make use of sophisticated generalizations (such as that about the proton-electron constitution of the hydrogen atom) which we certainly have not derived by simple enumeration of *instances*." [3] In such generalizations our inferences go beyond what has been directly presented in sensory experience, and also beyond what is directly confirmable by sensory experience. Now, it is assuredly possible to interpret these entities in a non-realistic (i. e., positivistic) manner, and thus not to regard what the scientist does as exemplifying transdiction. Such in fact would be Braithwaite's own view. However, as I shall now show, some scien-

[3] *Scientific Explanation*, p. 11.

tists, among whom we may number Boyle and Newton, have in fact regarded themselves as being able to refer to independently existing objects on the basis of inductive inferences which fall within the scope of scientific inquiry. Furthermore, both Boyle and Newton attempted to establish the validity of drawing such inferences. However, in doing so they appealed to what they regarded as general inductive principles: they did not consider themselves forced to defend epistemological realism as such. For them, science and epistemology did not in fact constitute two distinct sorts of disciplines which could be sharply distinguished from one another. They looked upon scientific inference as itself offering evidence for the existence of objects which are independent of sense perception, and as providing us with our most reliable knowledge of the characteristics possessed by those objects. How far we have since come from this position can be seen by briefly reverting to Kantian phenomenalism and to the direct realism which I have just mentioned. For the Kantian, science cannot describe transphenomenal entities; for the direct realist, inference to objects not directly experienced cannot rob us of that certainty which we possess in direct perceptual experience. In neither case is transdiction regarded as a function of science; in both, epistemology and science are independent of one another. Yet, if Professor Williams was right in his historical generalization, then scientists in the past have been concerned with problems which are of genuine epistemological significance, for they have attempted to discover and to state what exists independently of our experience. And perhaps it may be said that it is better to attempt to do this, if it can be done, than to argue *in abstracto* whether it is true or not true that there may be something which could so exist.

To this problem of the relation of the sciences to epistemology I shall ultimately return. However, in the present essay my task is primarily one of historical interpretation, and I wish to examine the realism which was inherent in the scientific and philosophic conceptions of Newton and of Boyle. For reasons of exposition, I shall reverse the natural, chronological order, and first discuss Newton.

II

Any belief that ordinary material objects are actually composed of atoms, and the acknowledgment that these atoms are not capable of being perceived by our senses, commits one to a belief in transdiction. That Newton believed in the real existence of atoms cannot be denied.[4] In his *Opticks* he even suggested that with better microscopes one might see the largest of these particles.[5] In the absence

[4] By "real existence" I mean that atoms exist independently of our thoughts, conceptions, or theories; that they are entities which have characteristics and stand in relations to one another independently of any relation which they may have to what we believe concerning them. To hold that atoms are only "constructs" which are used to render other data more intelligible, and that in employing the term "atom" we are only referring elliptically to such other data, would be to deny the "real existence" of atoms.

That Newton believed in the real existence of atoms (when that notion is taken in the above sense), seems to be recognized by all recent commentators on his views. For one of the clearest discussions of this point we may cite a conclusion reached by Marie Boas and Rupert Hall ("Newton's 'Mechanical Principles'"), who say, *inter alia*:

> Explicitly, then, it is for Newton impossible, *even in the case of gravitational attractions*, to step from the mathematical kind of reasoning to a physical kind of reasoning without introducing the corpuscular hypothesis. Even the astronomical sections of the *Principia* are by no means independent of the corpuscular conception (p. 176).

Furthermore, as we shall see, his atomism was by no means in conflict with the dictum "hypotheses non fingo" in the General Scholium of the *Principia*. Consequently the famous contrast which was drawn by Hume between Boyle and Newton cannot be regarded as justified. Hume said:

> Boyle was a great *partisan* of the mechanical philosophy; a theory which, by discovering some of the secrets of nature, and allowing us to imagine the rest, is so *agreeable* to the natural vanity and curiosity of men . . . . While Newton seemed to draw off the veil from some of the mysteries of nature, he showed at the same time the imperfections of the mechanical philosophy; and thereby restored her ultimate secrets to that obscurity in which they ever did and ever will remain. (*History of England*, VIII, 334.)

That Hume should have adopted this interpretation of Newton is not in itself surprising. However, it poses the problem of the extent to which such an interpretation was original with Hume, or by what stages it developed. Unfortunately, I am not in a position to suggest a reliable answer to that question.

[5] Bk. II, Part III, Prop. VII (p. 261).

For another equally clear expression of his atomism, see his essay of 1792,

of such experience, however, one may wonder why Newton was so thoroughly convinced of the truth of the corpuscular theory, and how he would justify the transdictive inferences which that conviction presumably involved.

To understand his conviction one must first take cognizance of the fact that the whole of modern science—in contrast to scholastic explanations which proceeded in terms of substantial and accidental forms—assumed the truth of a corpuscular theory of matter, and assumed that this theory could provide a means of explaining the ways in which material objects acted upon one another. As I Bernard Cohen remarks: "A basic 'corpuscular postulate' underlay all scientific thought in that age.—The 'new science' or 'new philosophy' which Newton called 'experimental philosophy' was simply a 'corpuscular philosophy.'" [6] Such a view was not, however, merely a tacit and unexamined presupposition in the thought of Newton; it was, rather, a perfectly explicit view which is to be found throughout his works.[7]

What formerly led some commentators to minimize Newton's belief in the real existence of atoms, and to neglect the role of transdictive inference to which this belief committed him, can be found in their interpretations of some of his methodological dicta,

entitled De Natura Acidorum, reprinted in Isaac Newton's Papers and Letters on Natural Philosophy, edited by I Bernard Cohen, pp. 256 ff. The essay is also to be found in The Correspondence of Isaac Newton, III, 205 ff.

Recently, in an article entitled "The Foundations of Newton's Philosophy of Nature," Richard Westfall has shown that Newton's student notebooks show the direct influence of Gassendi and of the other major atomists of the time.

[6] Franklin and Newton, p. 145.

[7] Cf. the following remark of A. Rupert Hall and Marie Boas Hall: "That a corpuscular or particulate theory was unreservedly adopted by him has long been abundantly evident from many passages in the Principia, and from the Quaeries in Opticks, to mention only discussions fully approved for publication by Newton himself. So far, then, Newton was undoubtedly a 'mechanical philosopher' in the spirit of the age, the spirit expressed, for example, by Boyle and Locke" ("Newton's Theory of Matter," p. 131).

In the Unpublished Scientific Papers of Isaac Newton, edited and translated by them, the same authors have a most informative account of Newton's theory of matter (pp. 183–213). For instances of Newton's atomism in the papers collected in that volume, I should especially cite pages 122, 306–7, 316–17, and 345.

and the emphasis which they have tended to place upon them. The most famous of these dicta is, of course, the familiar "hypotheses non fingo" contained in the general Scholium which Newton added to the *Principia* when he published the second edition of that work. Another such dictum, which is echoed in the same Scholium, is to be found in Newton's original preface to the *Principia*; it is usually quoted in the following form: "The whole burden of philosophy seems to consist in this—from the phenomena of motions to investigate the forces of nature, and from these forces to demonstrate other phenomena." This statement of principle has frequently been regarded as proving that Newton wished to restrict scientific explanation to the derivation of general laws from observation, and to the prediction of other observable phenomena on the basis of these laws. In other words, it is frequently claimed that Newton was *not* concerned with what might constitute the inherent nature of bodies, or the causes which explained their actions.

However, it is doubtful whether Newton's dictum should be interpreted in this restrictive, positivistic way. In the first place, a scrutiny of the dictum itself (totally apart from any questions raised by its context) shows that Newton held that the explanation of phenomena involved a discovery of the forces of nature, and not merely an extrapolation from past observations to future observations.[8] That this is a correct interpretation of the dictum, and that Newton was not averse to the notion of forces as explanatory concepts, may be made more evident if we now, in the second place, note what Newton's preface as a whole was attempting to show. If I am not mistaken, this preface constitutes nothing less than a defense of the view that the basic science of nature was mechanics, and Newton specifically defined rational mechanics as "the science of motions resulting from any forces whatsoever, and of the forces required to produce any motions, accurately proposed and demonstrated." The

---

[8] Newton used the terms "forces" and "powers" as equivalent terms in this preface, and there is no evidence in it that he was inclined to interpret the notion of a force (or a power) as synonymous with the fact that a body did actually move (or otherwise act) in a particular observable manner. In short, the notion of a force (or power) was not interpreted by him in the way in which it came to be interpreted by more positivistically inclined scientists and philosophers.

sarne emphasis on the notion of the forces which are to explain motions, becomes even more clear when Newton, in a passage immediately following his dictum, says that he is "induced by many reasons to suspect" that all of the phenomena of nature may depend upon "certain forces by which the particles of bodies, by some causes hitherto unknown, are either mutually impelled towards one another, and cohere in regular figures, or are repelled and recede from one another." And to this he added: "These forces being unknown, philosophers have hitherto attempted the search of Nature in vain." In the light of this commitment to mechanics as the basic discipline in the investigation of nature, we can see that the dictum with which we are here concerned should not be interpreted merely as a methodological dictum: it was also a statement of what might be called Newton's view of nature. For when he said that "the whole burden of philosophy seems to consist in this—from the phenomena of motions to investigate the forces of nature . . . ." we may take him to be saying that the forces of nature are to be investigated by investigating *motion*, and not by investigating the other observable qualities of things. Thus, I submit, the whole of this preface contains expressions of what are to be regarded as metaphysical commitments, and Newton's dictum should not be interpreted as if it were the expression of a positivistic theory of how science is to proceed.[9]

[9] Newton's commitments are evident in the opening sentence of the preface, for he identifies himself with those who "rejecting substantial forms and occult qualities, have endeavored to subject the phenomena of nature to the laws of mathematics." If it be thought that this does not express any specific ontological commitments, but is merely a methodological question, it is further to be noted that as he proceeds he brings out the difference between his views and the views of Descartes. He does so first by arguing that mechanics is more fundamental than geometry, and this argument is incompatible with Cartesian views on the relations of these sciences to each other, as well as being incompatible with Cartesian views of the nature of material substance. He does so again when, in the passages which we have already noted, he states his belief that all phenomena of nature are to be explained in terms of the attractive and repulsive forces inherent in the particles of bodies—a position opposed to the Cartesian principles of mechanical action. Thus, this preface clearly defines a specific ontological position: that of corpuscularianism in one of its special (non-Cartesian) forms. [On the meaning of "corpuscularianism," cf. below, pp. 88–90.]

Finally, in an attempt to interpret this frequently quoted dictum, we must note that even were it to be considered as a merely methodological dictum, it was not put forward by Newton as a description of the whole range of those methods which he himself either advocated or used. Actually, when we examine the context in which it appears, we find that Newton is probably only referring to the method which he had followed in Book III of the *Principia*. This seems probable in the light of what he says at the opening of Book III, where he clearly distinguishes between the mathematical method of the first two books and his aim in the third book. That passage reads:

> In the preceding books I have laid down the principles of philosophy; principles not philosophical but mathematical: such, namely, as we may build our reasonings upon in philosophical inquiries. These principles are the laws and conditions of certain motions, and powers or forces, which chiefly have respect to philosophy; but lest they should have appeared of themselves dry and barren, I have illustrated them here and there with some philosophical scholiums.[10]

In the light of this statement it would seem plausible to hold that in his Preface he is drawing exactly the same distinction between Books I and II on the one hand, and Book III on the other. Thus,

---

Precisely the same commitments are evident in Query 31 of the *Opticks* (p. 375 f.), as well as in the early manuscript, " On the Gravity and Equilibrium of Fluids," which—as we may incidentally note—contains a thorough-going criticism of Descartes. (For this manuscript, cf. *Unpublished Scientific Papers of Isaac Newton*, edited by A. R. and M. B. Hall, and especially page 122.)

[10] Cf. *Principia*, Cajori edition, p. 397.

Remembering that " philosophical " may perhaps best be interpreted as meaning " empirical, physical," it becomes clear that Newton was distinguishing between the mathematical demonstrations of the first two books and the empirical method of Book III. In fact, originally Book III was not even expounded in the same form as were the first two books, but was written in a more popular, nonmathematical way. (The Cajori edition of the *Principia* gives it in this form, after the conclusion of the final version of Book III.) Although Newton recast it " into the form of Propositions (the mathematical way) " in order to prevent disputes arising (cf. Cajori edition, p. 397), the above quotation shows that he did regard the method of treatment in the first two books as different from the method in the third.

this dictum should not be regarded as an attempt to explain the method which he was following in Books I and II; it cannot therefore be taken as an exhaustive statement of the Newtonian method.[11]

On the basis of all these considerations it would seem clear that the dictum which we have been discussing does not justify the view that Newton himself held to the restrictive and positivistic interpretation of scientific method which has been frequently attributed to him. To be sure, his reliance upon the notion of attractive forces as underlying the observable motions of bodies did open the way to the charge that he was guilty of reintroducing occult qualities into natural philosophy. It was as a means of defending himself against such a charge that in the General Scholium, which he added to the second edition of the *Principia*, he put forward the disclaimer: "hypotheses non fingo."

What constitutes a proper interpretation of this dictum is assuredly

---

[11] It will be well to quote the dictum in its context so that the reader may judge for himself whether it is designed to refer only to Book III, or whether it is to be taken as also applying to the mathematical methods of demonstration which Newton employed in laying the foundations for that concluding book. In the Cajori edition the passage reads:

. . . and therefore I offer this work as the mathematical principles of philosophy for the whole burden of philosophy seems to consist in this—from the phenomena of motions to investigate the forces of nature, and then from these forces to demonstrate the other phenomena; and to this end the general propositions in the first and second Books are directed. In the third Book, I give an example of this in the explication of the System of the World; for by the propositions mathematically demonstrated in the former Books, in the third I derive from the celestial phenomena the forces of gravity with which bodies tend to the sun and the several planets. Then from these forces, by other propositions which are also mathematical, I deduce the motions of the planets, the comets, the moon, and the sea.

In the original Latin preface it reads:

. . . et eâ propter, hæc nostra tanquam philosophiæ principia mathematica proponimus. Omnis enim philosophiæ difficultas in eo versari videtur, ut a phænomenis motuum investigemus vires naturæ, deinde ab his viribus demonstremus phænomena reliqua. Et huc spectant propositiones generales, quas libro primo et secundo pertractavimus. In libro autem tertio exemplum hujus rei proposuimus per explicationem systematis mundani. Ibi enim, ex phænomenis cœlestibus, per propositiones in libris prioribus mathematicè demonstratas, derivantur vires gravitatis, quibus corpora ad solem et planetas singulos tendunt. Deinde ex his viribus per propositiones etiam mathematicas, deducuntur motus planetarum, cometarum, lunæ et maris.

among the most vexed of all points in the history of modern scientific thought. On one matter, however, all recent commentators seem to agree: Newton was here speaking of the problem of explaining gravitation itself; he was not putting forward a general methodological principle.[12] What he wished to do was to fend off the question of how gravitation was to be explained; such a question, with which he did in fact elsewhere concern himself, was to form no part of the *Principia*.[13] It is also to be noted that the usual translation of Newton's dictum, "I frame no hypotheses" has been plausibly argued to be misleading: the dictum should rather be

[12] This was clearly stated by Cajori when he said: "In the first place, it should be noted that Newton does not advance 'hypotheses non fingo' as a general proposition, applying to all his scientific endeavor; it is used by him in a public statement relating to that special, that difficult and subtle subject, the real nature of gravitation" (*Newton's Mathematical Principles*, edited by Cajori, p. 671, note 55). I know of no recent commentator who would disagree with this statement; it is also worth noting that this opinion was shared by Philip E. M. Jourdain in his discussion of "hypotheses non fingo" in his series of interpretative articles on Newton's theory of the ether (cf. "Newton's Hypotheses of Ether and Gravitation," p. 250).

The recent commentators whose special discussions of this dictum (or of closely related topics) I should especially like to call to the reader's attention are the following, listed in chronological order:

(1) Cajori, as cited above, pp. 671–76; cf. also note 6, pp. 632–35.

(2) J. H. Randall, Jr., "Newton's Natural Philosophy," especially pp. 339–46.

(3) A. Koyré, "Pour une édition critique des œuvres de Newton."

(4) I Bernard Cohen, *Franklin and Newton*, pp. 125–45 and Appendix I.

(5) A. Koyré, L'hypothèse et l'expérience chez Newton."

(6) A. C. Crombie, "Newton's Conception of Scientific Method."

(7) E. W. Strong, "Hypotheses Non Fingo."

(8) A. Koyré, "Les Regulae Philosophandi."

(9) A. Rupert Hall and Marie Boas Hall, "Newton's Theory of Matter."

[13] He offered a speculative, mechanical account of gravitation in terms of the ether in his second paper on light and colors, and in his letter to Boyle. Furthermore, as can be seen in both the *Opticks* and his letters to Bentley, Newton felt that it *was* necessary to explain gravitation, and not take it as an innate quality of matter. However, what he insisted upon in the General Scholium is that he is *not* here attempting to explain gravitational phenomena *in any way*; he is therefore rejecting the charge that he is explaining them by means of an occult ("innate") quality. (On manifest vs. occult qualities, cf. also *Opticks*, Query 31, p. 401.)

translated "I *feign* no hypotheses."[14] And, finally, we must note that recent commentators all agree that in this passage Newton was using the term "hypotheses" in a special sense, distinct from the sense in which we, today, would use that term. What constitutes the precise sense in which he here spoke of "hypotheses" is by no means beyond debate.[15] However, in my opinion, our interpretation of Newton's precise meaning may perhaps best rely upon the letter which he wrote to Cotes on March 28, 1713. In that letter, which was written while the second edition of the *Principia* was going through the press, Newton instructed Cotes to add certain sentences to the General Scholium, and these sentences, he told Cotes, were for the purpose of "preventing exceptions against the use of the word 'Hypothesis.'"[16] The sentences which he added were the familiar ones which immediately follow the dictum "hypotheses non fingo." In the LaMotte translation they read: "for whatever is not deduced from the phenomena is to be called an hypothesis; and hypotheses, whether metaphysical or physical, whether of occult qualities or mechanical, have no place in experimental philosophy . . ." What is most enlightening in this letter to Cotes is Newton's explication of these added sentences. He wrote:

---

[14] Cf. Cohen, *Franklin and Newton*, p. 125 f. (note); also, Crombie, "Newton's Conception of Scientific Method," p. 8.

In support of this view it may be pointed out that "feign," not "frame," was the word used by Henry Jones when he translated Freind's defense of Newton against the Leibnizians. (Cf. Jones' abridgment of *The Philosophical Transactions of The Royal Society*, V, Part I, 428; the Latin version is to be found on page 330 of volume 27 of the *Transactions* themselves.)

[15] Crombie, in the article cited above, makes a claim more radical than that which seems to represent the consensus of opinion among other commentators. He regards Newton's use of the term "hypothesis" as being in this context equivalent to "fiction," or "hypothetical model," or "as-if construction" ("Newton's Conception of Scientific Method," p. 8). This interpretation receives some support from Cotes' discussion of hypotheses in his preface to the second edition of the *Principia* (Cajori ed., p. xxvii f.); it also receives support from some of the comparisons which Koyré makes between the use of the term "hypotheses" in earlier astronomy and Newton's use of it here (cf. Koyré, "Pour une édition critique . . . ," pp. 28 ff.). Nevertheless, I am inclined to take a more conservative view of the meaning of the term, as it is here used by Newton.

[16] The letter is to be found in Edleston, *Correspondence of Newton and Cotes*, pp. 154–56.

. . . As in Geometry the word Hypothesis is not taken in so large a sense as to include the Axiomes and Postulates, so in Experimental Philosophy it is not to be taken in so large a sense as to include the first Principles or Axiomes which I call the laws of motion. These principles are deduced from Phaenomena and made general by Induction: which is the highest evidence that a Proposition can have in this philosophy. And the word Hypothesis is here used by me to signify only such a Proposition as is not a Phaenomenon nor deduced from any Phaenomena but assumed or supposed without any experimental proof.

Basing our interpretation on this passage, we may say that what distinguishes "an Hypothesis" from legitimate explanatory propositions is the fact that the latter are either directly based upon observed phenomena or else are deduced from these phenomena.[17] In other words, Newton's test of the legitimacy of an explanatory proposition is, in the first instance, a question of how it has been arrived at, rather than being exclusively a question of how well it serves to explain further phenomena.[18] This emphasis upon the *derivation* of those explanatory principles which were to be used in

[17] It is to be noted that in his earlier controversies with Hooke and Pardies concerning the theory of colors, Newton had used the term "hypotheses" in a slightly different way: hypotheses were explanatory principles which *might be* deduced from phenomena. (Cf. especially four of Newton's letters to Oldenburg, given in H. W. Turnbull; *The Correspondence of Isaac Newton*, I, and there numbered 66, 67, 75, and 146.) However, in these debates Newton sought to avoid introducing any hypotheses even in this sense, since, as he said, he could think of several alternative hypotheses each of which would be compatible with all of the phenomena. (Cf. Letter 67, p. 174. Also, cf. Letter 66, p. 169.) However, when engaged in the debate with Hooke, Newton did formulate his argument in terms of an hypothesis, for the sake of exposition; however, he warned the reader that because of this fact he wished that "no man may confound this with my other discourses, or measure the certainty of one by the other" (Letter 146, p. 364). [This letter is better known as "Newton's Second Paper on Light and Colours."]

[18] I do not wish to be taken as minimizing the emphasis which Newton placed upon the application of laws as explanations of further cases: as we have noted, it was his aim to "demonstrate other phenomena" on the basis of those forces which he sought to discover. However, there is a notable contrast between Newton's emphasis on the derivation of an explanatory theory, and the emphasis which is currently placed on *confirmation* (and, especially, confirmation through predictive power), regardless of the origin of the theory.

experimental philosophy can be seen in other passages, and this emphasis is in fact the main burden of the lines which Newton instructed Cotes to add to the General Scholium, and which we may now quote again, but in full:

. . . for whatever is not deduced from the phenomena is to be called an hypothesis; and hypotheses, whether metaphysical or physical, whether of occult qualities or mechanical, have no place in experimental philosophy. In this philosophy particular propositions are inferred from the phenomena, and afterwards rendered general by induction. Thus it was that the impenetrability, the mobility, and the impulsive force of bodies, and of the laws of motion and of gravitation, were discovered.

In short, it was because of their derivation, and not merely because of their applicability in explaining phenomena, that Newton claimed that his laws were not "hypotheses," in the pejorative sense in which he was using that term.[19]

That Newton was conscious of this difference between an explanatory theory which had been derived from phenomena, and those hypotheses (whether metaphysical or physical) which were not so derived, but which could nonetheless explain phenomena, seems to me to be quite clear from the fourth "Rule of Reasoning in Philosophy" which he inserted in the third edition of the *Principia*. This rule reads:

---

[19] That this is the correct interpretation of Newton's usage is also attested by a second letter which Newton wrote to Cotes on the same subject during the course of the following week:

On Saturday last I wrote to you, representing that Experimental philosophy proceeds only upon Phenomena and deduces general Propositions from them only by Induction. And such is the proof of mutual attraction. And the arguments for ye impenetrability, mobility and force of all bodies and for the laws of motion are no better. And he that in experimental Philosophy would except against any of these must draw his objection from some experiment or phaenomenon and not from a mere Hypothesis, if the Induction be of any force. (Edleston, *Correspondence of Newton and Cotes*, p. 156.)

Further confirmation of my interpretation seems to me to be found in the way in which Cotes uses the term "hypothesis" in his preface to the second edition of the *Principia*. (Cf. especially the paragraph starting at the foot of page xx of the Cajori edition; cf. also that starting at the foot of page xxvi which deals with the same question, though it does not use the word "hypothesis.")

In experimental philosophy we are to look upon propositions inferred by general induction from phenomena as accurately or very nearly true, notwithstanding any contrary hypotheses that may be imagined, till such time as other phenomena occur, by which they may either be made more accurate, or liable to exceptions.

This rule we must follow, that the argument of induction may not be evaded by hypotheses.[20]

Similarly, in concluding the *Opticks*, Newton's rejection of hypotheses is couched in terms of their lack of empirical *derivation*, not in terms of any lack of explanatory power. He there says:

. . . The method of analysis ought ever to precede the method of composition. This analysis consists in making experiments and observations, and in drawing general conclusions from them by induction, and admitting of no objections against the conclusions, but such as are taken from experiment, or other certain truths. For hypotheses are not to be regarded in experimental philosophy.[21]

And in this connection one final, perfectly explicit passage of the same sort may be cited from one of Newton's letters to Oldenburg:

In the meane while give me leave to insinuate that I cannot think it effectuall for determining truth to examin the severall ways by wch Phaenomena may be explained, unless where there can be perfect enumeration of all those ways. You know the proper Method for inquiring after the properties of things is to deduce them from Experiments. And I told you that the Theory wch I propounded was evinced to me, *not by inferring tis thus because not otherwise*, that is not by deducing it onely from a confutation of contrary suppositions, but *by deriving it from Experiments concluding positively & directly*.[22]

To be sure, in the light of more recent discussions of scientific method it is not easy to see just how it is possible to "conclude positively and directly" from any set of experiments to a general

---

[20] Cf. Cajori edition, p. 400.
[21] Newton, *Opticks*, p. 404.
[22] Letter 75, dated 6 July 1672, in Turnbull, *Correspondence of Newton*, I, 209.

theory. Nor is it clear why in the famous "hypotheses non fingo" passage, and elsewhere, Newton should speak of "deducing" general propositions from observed phenomena. However, as William Kneale has pointed out, what seems to be at issue is simply Newton's desire to insist that the general principles be closely tied to the evidence from which they were derived.[23] Thus, once again we see that what constitutes "an hypothesis," as distinct from a legitimate scientific explanation, becomes for Newton primarily a question of how that general proposition has been derived.

Turning now to the question of confirmation, we see that Newton seeks for confirmatory evidence for his theories by "rendering them general by induction," a phrase which he frequently repeats.[24] We may, I suppose, take it for granted that by "induction" Newton is referring to the application of his theory to further particular cases in order to test its adequacy. This would be a normal construction of his use of the term not only in this passage, and in Rule IV, but in Query 31 of the *Opticks* as well. However, what seems not to have been explicitly noted by those who have discussed Newton's philosophy of science is the weight which he apparently attached to the notion of rendering his theories *general* by means of induction. Generality could mean—and does mean for Newton—that no exceptions to the empirically derived propositions are to be found when one examines further cases of the same sort.[25] This, however, seems

[23] Speaking of the passage from the General Scholium in which Newton says that in the experimental philosophy "propositions are deduced from phenomena and rendered general by induction," Kneale says:

> It will be noticed that Newton speaks in a very curious way of *deducing* propositions from phenomena. This expression occurs in other places, and we must assume that Newton used it deliberately; but it obviously cannot mean what is ordinarily called deduction, and I can only conclude that Newton meant that the propositions which interested him were derived from observation in a very strict way. (*Probability and Induction*, cf. pp. 98–101.)

That this is correct is suggested by the passage from the letter to Oldenburg which I have just quoted.

[24] For example, not only in the passage in the General Scholium which follows immediately upon the dictum "hypotheses non fingo," but also in Query 31 in the *Opticks*, in the two letters to Cotes which we have cited, and in Rule IV, which has been cited above.

[25] For example, in Query 31 of the *Opticks* he says: "And if no exception

to me to constitute only its minimal meaning for Newton. In addition, he sought "generality" in a stronger sense; namely, in that sense in which something which has been "rendered *general* by induction" has been shown to apply not only to all cases "of the same type," but to apply universally. In other words, in my opinion, Newton was holding that what is truly "general" is what is applicable throughout nature. Another way of expressing Newton's meaning is to say that he was convinced not only of the *invariance* of those laws which the experimental philosophy had discovered, but he was convinced of their *universal applicability* as well. And, actually, it was in their universal applicability that Newton found their confirmation; it was for this reason that he was willing to accept them even in the absence of any knowledge concerning their causes.[26] If we ask why he was so certain that universality, taken in this sense, guaranteed the truth of empirically established propositions, the answer seems to lie in his acceptance of the proposition that Nature always acts in the same manner.[27] Granted this assump-

---

occur from phaenomena, the conclusion may be pronounced generally. But if at any time afterwards any exception shall occur from experiments, it may then begin to be pronounced with such exceptions as occur " (p. 404).

[26] Cf. the following passage from Query 31 of the *Opticks*, noting also how the specificity of the so-called occult qualities of the Aristotelians was taken as counting against them:

To tell us that every species of things is endow'd with an occult specifick quality by which it acts and produces manifest effects, is to tell us nothing: but to derive two or three principles of motion from phaenomena, and afterwards to tell us how the properties and actions of all corporeal things follow from those manifest principles, would be a very great step in philosophy, though the causes of those principles were not yet discover'd: and therefore I scruple not to propose the principles of motion abovemention'd, they being of very general extent, and leave their causes to be found out (p. 401 f.).

[27] It is to be noted that in two of the passages in which induction is discussed, Newton introduces a proposition of this sort, almost as if it were an axiom. Cf. Query 31 of the *Opticks*, where he says, "Nature is very consonant and conformable to herself" (p. 376); also, in Rule III of Book III of the *Principia*, he says, "Nature, which is wont to be simple, and always consonant to itself . . ." (p. 398 f.).

Thanks to the publication of previously unpublished manuscripts from the Portsmouth Collection by A. R. and M. B. Hall, two further passages may be added in which Newton—in drafts of the Preface and the Conclusion—speaks

tion, whatever has not only been derived from observation, but has been shown to be capable of being rendered *general* by induction, may be taken as true.

Having reached this point, we are now, I believe, in a position to return to our original question [28] and ask on what basis Newton could justify his belief in the real existence of atoms, and on what basis he could justify the transdictive inferences which that conviction presumably involved. If Newton were to be interpreted as holding a positivistic theory of scientific inference, this question would have admitted of no satisfactory answer: we should have had to charge Newton with a fundamental inconsistency in his thought. However, having seen that Newton did in fact seek causal explanations of phenomena, and that he was not interpreting causation in terms of directly observed sequences, we need not be surprised to find him using the corpuscular theory as an explanatory notion basic to his experimental philosophy, so long as this theory was not "an hypothesis" in his pejorative sense of that term. Now, as we have seen, to escape being "an hypothesis" the corpuscular theory would have to have been derived from the observation of phenomena. Furthermore, as we have seen, before being accepted as true, any theory must be capable of being rendered general by induction. One might readily wonder how either of these two stipulations of the Newtonian method could have been fulfilled in the case of the corpuscular theory. However, if one turns to Newton's "Rules of Reasoning in Philosophy" and examines Rule III as it is stated in the second edition of the *Principia*,[29] one finds a perfectly explicit statement of the justificatory principle which Newton could use in answering any challenge on either of these counts. I am not aware

of Nature as being "simple" and "conformable to itself": cf. *Unpublished Scientific Papers of Isaac Newton*, pp. 307, 333.

[28] Cf. above, pp. 66 f.

[29] The last three sentences of Rule III were added in the third edition, and are obviously designed to meet the objection that Newton believed gravity to be an "essential" and "innate" quality of matter. Rule IV, in its present form, was also not added until the third edition.

On the changes in this section of the *Principia*, cf. Cohen, *Franklin and Newton*, pp. 584–85; also "Pour une édition critique des œuvres de Newton" and "Les Regulae Philosophandi" by Koyré.

that this significance of Rule III for Newton's atomism has been pointed out before, and I shall therefore proceed to argue it with some care.

By and large, the four "Rules of Reasoning in Philosophy," as given at the opening of Book III in the final edition of the *Principia*, would seem to be concerned only with general methodological principles of explanation, and they would not appear to contain any very surprising tenets. For example, if one examines only Rules I and II one finds oneself on thoroughly familiar ground. Rule I states:

We are to admit no more causes of natural things than such as are both true and sufficient to explain their appearances.

To this purpose the philosophers say that nature does nothing in vain, and more is in vain when less will serve; for Nature is pleased with simplicity, and affects not the pomp of superfluous causes.

And Rule II states:

Therefore to the same natural effects we must, so far as possible, assign the same causes.

As to respiration in a man and in a beast; the descent of stones in *Europe* and in *America*; the light of our culinary fire and of the sun; the reflection of light in the earth, and in the planets.

These two rules are clearly methodological rules of a familiar sort: the first is the rule of simplicity, the second is a methodological version of the principle of the uniformity of nature.[30] And Rule IV (which we have previously quoted) would, when taken by itself, also seem to be a purely methodological principle concerned with our right to accept any inductively established generalization, to which there was added a warning against "feigning" hypotheses. Rule IV reads:

In experimental philosophy we are to look upon propositions inferred by general induction from phenomena as accurately or very

---

[30] This is also the opinion of R. M. Blake in an article worth citing: "Sir Isaac Newton's Philosophy of Scientific Method," p. 485. However, Blake regards Rule III as being "no more than a reformulation of the first two, with reference to another set of conclusions"; in this he seems to me to be in error.

*nearly true, notwithstanding any contrary hypotheses that may
be imagined, till such time as other phenomena occur, by which
they may either be made more accurate, or liable to exceptions.*

This rule we must follow, that the argument of induction may
not be evaded by hypotheses.

However, when *this* Rule is taken as an addition to the lengthier
and more important Rule III, it takes on an added significance.
In its insistence on accepting what has been derived from a consideration of phenomena, and rejecting what is merely compatible with
those phenomena, it constitutes a supplement to the aim of Rule
III, which, as we shall now see, states Newton's theory of the basis
and proper course of all scientific reasoning.

In Rule III Newton holds, I suggest, that all scientific reasoning
has its basis in sense experience; that such reasoning must seek the
universal characteristics in the phenomena with which it deals; and
that any characteristics common to all phenomena which we have
experienced may be predicated of all objects whatsoever, whether or
not these have been experienced, and whether or not they are
experienceable. Thus, this is a rule which recommends two things
which are often thought to be incompatible. On the one hand it
endorses the view that all knowledge must start from sense experience
and must proceed by forming generalizations which are based upon
the repeated traits found within experience. On the other hand, it
states that it is legitimate to use these generalizations not only when
dealing with phenomena which are in principle confirmable within
experience, but that it is also proper to use them to arrive inductively
at generalizations concerning what transcends entities which are
experienceable.

In quoting from the explanation which is added to Rule III, I
shall italicize those passages which refer to the legitimacy of general-
izing to what transcends experience.

Rule III

*The qualities of bodies, which admit neither intensification nor
remission of degrees, and which are found to belong to all bodies*

within the reach of our experiments, are to be esteemed the universal qualities of all bodies whatsoever.

For since the qualities of bodies are only known to us by experiments, we are to hold for universal all such as universally agree with experiments; and such as are not liable to diminution can never be quite taken away. We are certainly not to relinquish the evidence of experiments for the sake of dreams and vain fictions of our own devising; nor are we to recede from the analogy of Nature, which is wont to be simple, and always consonant to itself. We no other way know the extension of bodies than by our senses, *nor do these reach it in all bodies; but because we perceive extension in all that are sensible, therefore we ascribe it universally to all others also.* That abundance of bodies are hard, we learn by experience; and because the hardness of the whole arises from the hardness of the parts, *we therefore justly infer the hardness of the undivided particles not only of the bodies we feel but of all others.* That all bodies are impenetrable, we gather not from reason, but from sensation. The bodies which we handle we find impenetrable, *and thence conclude impenetrability to be an universal property of all bodies whatsoever. That all bodies are movable, and endowed with certain powers (which we call the inertia) of persevering in their motion, or in their rest, we only infer from the like properties observed in the bodies which we have seen.* The extension, hardness, impenetrability, mobility, and inertia of the whole, result from the extension, hardness, impenetrability, mobility, and inertia of the parts; and hence *we conclude the least particles of all bodies to be also all extended, and hard and impenetrable, and movable, and endowed with their proper inertia.* And this is the foundation of all philosophy.[31]

The final paragraph of Rule III, which I shall now quote, explains by implication why Newton found it necessary to formulate this

---

[31] Cf. *Opticks*, Query 31 (p. 404) where Newton says: "And although the arguing from Experiments and Observations by Induction be no Demonstration of general Conclusions; yet it is the best way of arguing which the Nature of Things admits of, and may be looked upon as so much the stronger, by how much the Induction is more general. And if no Exception occur from Phaenomena, the Conclusion may be pronounced generally." It is my contention that this last sentence, like Rule III, is to be interpreted as holding even beyond the realm of "Phaenomena," i. e., beyond the realm of *directly* observable events.

rule: he was forced to do so if he were to justify holding that gravitational force not only explained the behavior of such molar phenomena as the planetary orbits and the behavior of the tides, but that it did so because gravitational force operated between the component particles of all molar matter.[32] This paragraph reads:

Lastly, if it universally appears, by experiments and astronomical observations, that all bodies about the earth gravitate towards the earth, and that in proportion to the quantity of matter which they severally contain; that the moon likewise, according to the quantity of its matter, gravitates towards the earth; that, on the other hand, our sea gravitates towards the moon; and all the planets one towards another; and the comets in like manner towards the sun; we must, in consequence of this rule, universally allow that all bodies whatsoever are endowed with a principle of mutual gravitation. For the argument from the appearances concludes with more force for the universal gravitation of all bodies than for their impenetrability; of which, among those in the celestial regions, we have no experiments, nor any manner of observation.

What this rule thus advocates is the validity of using data within experience in order to make inferences to data which not only have not been experienced, but which cannot be experienced, and this is claimed to be valid *so long as the latter can be assumed to be of the same kind as those found within experience, and this assumption can legitimately be made whenever we are dealing with characteristics which are found to hold without exception within our experience.* This, I submit, is a very special use of the principle of the uniformity

[32] This, as we have seen, is his view as stated in the preface to the 1st edition. In the General Scholium added in the 2nd edition Newton says: "Gravitation towards the sun is made up out of the gravitations towards the several particles of which the body of the sun is composed." (Cajori ed. p. 546.) This, be it noted, is said just before Newton pronounces his famous dictum "hypotheses non fingo." No clearer proof is needed that his dictum merely refers to the fact that he is not attempting to frame a hypothesis concerning the cause of gravitational forces within the limits of the *Principia*, i. e., he is not invoking an occult quality, nor giving any mechanical explanation, such as the Cartesian, for the causes of that force whose operation he has derived from his consideration of phenomena.

of nature. Instead of being used merely to justify such maxims as "Given the same effects we can assume the same causes," or "Given the same causes, the same effects will follow," the idea of the uniformity of nature is extended to justify a maxim of the following sort: "Characteristics which are invariably associated with experienced objects are also characteristics of all objects whatsoever."

The latter maxim, I now submit, would be precisely the sort needed to justify a distinction between those qualities which are to be called "primary" and those qualities which are to be called "secondary." And it is to be noted that in Rule III the precise qualities which Newton designates as those which do not admit of either "intensification nor remission of degrees, and which are found to belong to all bodies within the reach of our experiments," are extension, hardness, impenetrability, mobility, and inertia, i. e., they are those which are generally included among the primary qualities.[33] With respect to divisibility, Newton explicitly discusses it in a passage which I did not previously quote. He was unwilling to state whether divisibility was a universal characteristic of matter since experience did not in fact show that all matter was divisible.[34] This cautious approach to the question of divisibility is almost surely a reflection of Newton's own belief in indivisible atoms, which finds expression in the Opticks when he says:

All these things being consider'd, it seems probable to me, that

---

[33] I am not certain of the reason for Newton's separation of hardness from impenetrability. This separation is also to be found in Opticks, Query 31 (p. 389). However, my colleague Richard T. Cox has very plausibly suggested to me that by "hardness" Newton probably meant "not-penetrable and not-deformable," whereas "impenetrability" does not exclude deformability.

[34] As he says:

Moreover that the divided but contiguous particles of bodies may be separated from one another, is matter of observation; and, in the particles that remain undivided, our minds are able to distinguish yet lesser parts, as is mathematically demonstrated. But whether the parts so distinguished, and not yet divided, may, by the powers of Nature, be actually divided and separated from one another, we cannot certainly determine. Yet, had we the proof of but one experiment that any divided particle, in breaking a hard and solid body, suffered a division, we might by virtue of this rule conclude that the undivided as well as the divided particles may be divided and actually separated to infinity.

God in the Beginning form'd Matter in solid, massy, hard, impenetrable movable Particles, of such Sizes and Figures, and with such Properties, and in such Proportion to Space, as most conduced to the End for which he form'd them; and that these primitive Particles being Solids, are incomparably harder than any porous Bodies compounded of them; even so very hard, as never to wear or break in pieces; no ordinary Power being able to divide what God himself made one in the first Creation.[35]

With respect to such qualities as color, taste, odor, and the like, it is clear that Rule III would not justify our attributing them to the atoms, for many objects appear tasteless and odorless, and some objects—both air and glass—are without color. Thus Newton used Rule III as a means of distinguishing between primary and secondary qualities, and of justifying his belief as to which qualities were really the primary qualities of bodies. I am not aware that this has previously been pointed out.[36]

That the foregoing is a correct interpretation of Newton's philosophy of science may perhaps be made even more plausible by noting certain passages in the Scholium which follows the definitions given at the outset of the *Principia*. This Scholium opens with the following sentences:

Hitherto I have laid down the definitions of such words as are less known, and explained the sense in which I would have them understood in the following discourse. I do not define time, space, place, and motion, as being well known to all. Only I must ob-

---

[35] *Opticks*, Query 31 (p. 400). Thus Newton rejected the Cartesian form of the corpuscular philosophy in favor of a true atomism.

[36] Strangely enough, Rule III has not been much discussed, though its length and the changes between the first and second editions might have called attention to it. For example, though Burtt cites it twice (*Metaphysical Foundations of Modern Physical Science*, pp. 218-19 and 231-32) he cites it only as a restatement of the simplicity and uniformity of nature. Koyré does discuss it in "Pour une édition critique des œuvres de Newton" as well as in "Les Regulae Philosophandi" (as cited in note 10, above), but he does not show its methodological significance. The nearest approach to doing so is, I believe, to be found in K. Lasswitz: *Geschichte der Atomistik vom Mittelalter bis Newton*, II, 557-58. Lasswitz specifically points out that Newton is trying to derive all knowledge from sense experience; he also adds: "Das ist freilich Lockes Standpunkt." (See my discussion below.)

serve, that the common people conceive those quantities under no other notions but from the relation they bear to sensible objects. And thence arise certain prejudices, for the removing of which it will be convenient to distinguish them into absolute and relative, true and apparent, mathematical and common.[37]

There follow the famous discussions of absolute vs. relative time, space, place, and motion. With these discussions we shall not be concerned. What is to be noted is that Newton has in effect insisted that it is through sense experience that we come in the first instance to understand time, space, place, and motion, and as Léon Bloch has pointed out, he also assumes that it is through sense experience that we originally learn what is to be meant by solidity.[38] However, it will have been noted that in this passage Newton also insists that we must not rely upon sense experience to give us a true conception of the actual natures of those quantities with which sense first acquaints us. As he later points out in the same Scholium, in common affairs we use relative places and motions instead of absolute ones, and this suffices for our ordinary purposes, "but in philosophical disquisitions, we ought to abstract from our senses, and consider things themselves, distinct from what are only sensible measures of them."[39] As he then admits concerning absolute space, "the parts of that immovable space, in which those motions are performed, do by no means come under the observation of our senses."[40] Thus Newton holds that in scientific inquiry the senses stand in need of correction through arguments drawn from mental experiments (for it is thus that he proceeds in this Scholium), and through rational generalizations of that which we discover through the senses. Yet, as is evident from his constant rejection of "occult qualities" in

[37] Principia, p. 6.
[38] L. Bloch, La philosophie de Newton, pp. 135 ff. In what immediately follows I am in agreement with Bloch's interpretation. However, his interpretation of Newton's thought as a whole makes a positivistic philosopher of Newton (cf., for example, p. 332), and with this I am of course not in agreement.
[39] Principia, p. 8.
[40] Ibid., p. 12.

favor of "manifest qualities," [41] and in his insistence on what can be deduced from "Phaenomena," all of our knowledge must be ultimately derived from sense experience. That there is no contradiction between these two positions follows from Rule III: whatever we find to be universally characteristic of that which our senses reveal we can generalize and hold to be true even beyond the limits of sense.[42]

Had Rule III in its present form been in the first edition of the *Principia*, we might suspect that John Locke's treatment of the primary qualities was not wholly independent of it, for this Rule

[41] The contrast between occult and manifest qualities is most sharply drawn in Query 31 of the *Opticks*, p. 401. In the *Principia* the opening sentence of Newton's preface to the first edition stresses his rejection of occult qualities and substantial forms.

[42] That this is the correct interpretation of the role of Rule III in Newton's conception of method, and that I am by no means placing too much emphasis upon it, may perhaps be suggested by Henry Pemberton's explication of it in *A View of Sir Isaac Newton's Philosophy*, p. 25. After stating the rule in brief, Pemberton says:

In this precept is founded that method of arguing by induction without which no progress could be made in natural philosophy. For as the qualities of bodies become known to us by experiments only; we have no other way of finding the properties of such bodies, as are out of our reach to experiment upon, but by drawing conclusions from those which fall under our examination.

Thus, Pemberton stresses the role of sense experience as the foundation of all knowledge. However, he too stresses the role of generalization beyond what can be given through sensory observation, and he does so by means of an example which is not found in Newton's own explication of Rule III. This illustration is the fact that we do attribute extension to the fixed stars on the basis of this rule, rather than on the basis of direct observation, for—as he points out—" the more perfect our instruments are, whereby we attempt to find their visible magnitude, the less they appear; insomuch that all the sensible magnitude, which we observe in them, seems only to be an optical deception by the scattering of their light." Thus, the senses stand in need of correction in specific cases, even though they provide the ultimate source of all scientific knowledge. Rule III is taken as serving as the basis for distinguishing between those observed qualities which do reside in nature independently of us, and those which do not.

My interpretation is also compatible with the interpretation offered by F. Rosenberger, *Isaac Newton und seine physikalischen Prinzipien*, although his discussion lays less stress upon the epistemological issue which Rule III was presumably able to solve.

would have served as an ideal justification of what, in the preceding chapter, we have seen that Locke was seeking to hold. Furthermore, Newton's appeal to sense experience as the *foundation* of all of our knowledge, but *not* as the criterion of its truth, is also reminiscent of Locke. Yet, as in the case of Locke, we may wonder whether in fact Newton's method of proceeding—his attempt "from the phenomena of motions to investigate the forces of nature, and then from these forces to demonstrate the other phenomena"[43]—would have sufficed to establish the existence of atoms, whose existence his natural philosophy did in fact presuppose. As we shall later note, Rule III was not a sufficiently powerful methodological principle for this purpose. However, it was not Newton's purpose, any more than it had been Locke's to *establish* the truth of atomism: each took it as established and simply utilized it. The function of Rule III was, then, simply that of showing that its acceptance was consonant with the methodological principles of the new philosophy.[44]

III

It is to Boyle, rather than to Newton or Locke, to whom we must look if we are to examine an attempt to establish the superiority of corpuscular philosophy over its chief alternatives. Boyle's concern with this problem as contrasted with Newton's indifference to it, is in part explicable by the fact that Boyle, writing some twenty-five years earlier,[45] had more opponents of the new philosophy to face. How-

[43] Preface to the first edition of the *Principia*, p. xvii f.

[44] It is to be noted that, according to A. Rupert Hall and Marie Boas Hall (cf. *Unpublished Scientific Papers of Isaac Newton*, p. 187), Newton had from his student days been acquainted with Boyle's *Origin of Forms and Qualities*, which contained the basic arguments which Boyle used to establish the truth of his corpuscularianism. I have found only one passage in which Newton, like Boyle, gives empirical arguments for a corpuscular theory of the structure of matter (cf. *Unpublished Scientific Papers of Isaac Newton*, pp. 316–17): usually his arguments presuppose the truth of atomism and simply serve to explain specific effects in terms of this presupposition (e. g., Hall and Hall, *Unpublished Scientific Papers*, pp. 345–46).

[45] Boyle's *Spring and Weight of the Air* dates from 1660, *Certain Physio-*

ever, an even more important reason is to be found in the fact that the problems with which Boyle was dealing were not primarily problems in mechanics, and it was therefore still customary to invoke "occult qualities" and "forms" to explain them. For these two reasons Boyle was not able to take the new corpuscular philosophy for granted, but found it necessary first to define wherein it differed from other positions, after which he was forced in work after work to prove the advantages which it had over the views of both the peripatetics and the chemists.

Before following Boyle's attempts to establish the corpuscular philosophy in all of the fields that were of interest to him, it will be necessary to consider and reject one rather widely held interpretation of his thought. This interpretation regards him as holding a positivistic philosophy of science, and as "refusing to take his corpuscular and mechanical philosophy in a metaphysical sense, that is, as anything more than the most simple and fruitful hypothesis known to him." [46] Were this interpretation to be accepted, one could not regard Boyle as believing in the possibility of "transdiction," and this in turn, as we shall see, would make it difficult to know how to interpret a great part of his scientific work. It is therefore necessary to attempt to refute at the outset the view that Boyle is to be classed as holding a merely heuristic view of atomism.

In what is perhaps the most explicit of the supposedly positivistic

logical Essays was published in 1661, and his Origin of Forms and Qualities in 1666; whereas Newton's Principia was published in 1687.

[46] Philip P. Wiener, in "The Experimental Philosophy of Robert Boyle," p. 605. Cf. "Locke avait appris à Oxford de Ward, de Wallis, ou de Boyle, par leur enseignement ou par leurs ouvrages, que l'essence des choses ou les véritables causes des phénomènes nous sont inconnues," H. Ollian, La philosophie générale de Locke, p. 149; cf. also p. 111. J. W. Yolton also speaks of "Boyle's phenomenalistic doctrine . . . concerning substance," John Locke and the Way of Ideas, p. 126 and Leroy E. Loemker shares the same view (cf. "Boyle and Leibniz," p. 31).

On the other hand, in his dissertation Robert Boyles Naturphilosophie, apparently written under the direction of von Hertling, the Locke scholar, J. Meier rejects the notion that the skeptical strain was a fundamental characteristic in the thought of Boyle. Nor does M. Giua's recent study of Boyle suggest that he would share Wiener's view. However, that view is sufficiently important to demand examination.

passages, Boyle links the atomistic and Cartesian views of matter, contrasting them with the Aristotelian view. He characterizes both as examples of the "Corpuscularian philosophy," [47] but after noting some of the many differences between them, he refuses to take sides with either the Cartesians or the atomists.[48] Now, the fact that Boyle was willing not to enter into controversy concerning the topics in which the two branches of the corpuscular philosophy were opposed might be taken to suggest that he had in fact no serious interest in the details of their explanations of phenomena, and from this it might then be supposed that he was only interested in employing the notion of corpuscles in so far as such a notion was a useful heuristic device. This interpretation might be further fortified by noting other passages in which Boyle explicitly held that when we invoke corpuscles to explain a particular effect we are often unable to say with accuracy just what these corpuscles must be like. For example, in one passage he explicitly held that all that the ancient atomists had succeeded in doing was to show that effects *might* have been produced from certain causes; they were unable to prove that these effects had been thus caused.[49] As he remarked in illustration of this point:

For as an artificer can set all the wheels of a clock agoing as well

---

[47] Boyle apparently coined this term himself: cf. *Works*, III, 5.

[48] This often quoted passage reads as follows:

I know that these two sects of modern naturalists [i. e., the Cartesians and the atomists] disagree about the notion of body in general, and consequently about the possibility of a true vacuum; as also about the origin of motion, the infinite divisibleness of matter, and some other points of less importance than these: but in regard that some of them seem to be rather metaphysical than physiological notions, and that some others seem rather to be requisite to the explication of the first origin of the universe, than of the phaenomena of it, in the state wherein we now find it; in regard to these, I say, and some other considerations, and especially for this reason, that both parties agree in deducing all the phaenomena of nature from matter and local motion; I esteemed that, notwithstanding these things, wherein the Atomists and Cartesians differed, they might be thought to agree in the main, and their hypotheses might by a person of reconciling disposition be looked on as, upon the matter, one philosophy. Which, because it explicates things by corpuscles, or minute bodies, may (not very unfitly) be called corpuscular. (*Ibid.*, I, 355–56.) *Cf.* also, *ibid.*, III, 7.

[49] *Ibid.*, II, 45–46.

with springs as with weights . . . so the same effects may be produced by divers causes different from one another; and it will oftentimes be very difficult, if not impossible, for our dim reasons to discern surely, which of these several ways, whereby it is possible for nature to produce the same phaenomena, she has really made use of to exhibit them.

In short, as he says in this passage, "it is a very easy mistake for men to conclude, that because an effect may be produced by such determinate causes, it must be so, or actually is so." [50] However, this statement must be treated with caution, for Boyle called attention to the very same fact when he was explaining the spring of the air. In that discussion he was willing to suggest two alternative hypotheses to account for the phenomena: one involved his own analogy of the springiness of fleece, the other was the explanation given by the Cartesian system. As he said:

By these two differing ways, my Lord, may the springs of the air be explicated. But though the former be that, which by reason of its seeming somewhat more easy, I shall for the most part make use of in the following discourse; yet am I not willing to declare peremptorily for either of them against the other. And indeed, though I have in another treatise endeavoured to make it probable, that the returning of elastical bodies (if I may so call them) forcibly bent, to their former position, may be mechanically explicated; yet I must confess, that to determine whether the motion of restitution in bodies proceed from this, that the parts of a body of a peculiar structure are put into motion by the bending of the spring, or from the endeavour of some subtle ambient body . . . : to determine this, I say, seems to me a matter of more difficulty, than at first sight one would easily imagine it. Wherefore I shall decline meddling with a subject, which is much more hard to be explicated than necessary to be so by him, whose business it is not, in this letter, to assign the adequate cause of the spring of the air, but only to manifest, that the air hath a spring, and to relate some of its effects. [51]

---

[50] Locke also uses the clock illustration (*Essay*, Bk. III, Ch. VI, Sec. 39). An obvious source of this illustration is to be found in Descartes, *Principles of Philosophy*, Pt. IV, Prop. CCIV.

[51] *Works*, I, 12.

This, in all conscience, sounds positivistic enough. However, it is to be noted that Boyle was simply avoiding the issue *in that particular place*, for he was ready to grant that it was a problem of considerable difficulty. It was, nevertheless, a problem to which he returned in his *First Continuation of New Experiments Physico-Mechanical Touching the Spring of the Air*, where he propounded reasons for rejecting the Cartesian theory.[52] In this case, as in others, Boyle is proceeding piecemeal, and is limiting himself to a specific type of problem and setting out his theory with a particular type of opponent in mind.[53] What we may say is that Boyle was inclined to reject debates over the specific characteristics of the nonobservable elements of things when these debates could not be settled by inferences drawn from observations, and especially when such debates were motivated by an interest in system-building. Like Bacon, whom he so greatly admired, Boyle was averse to the formation of systems, and contrasted the systematic method of procedure with the experimental.[54] The corpuscular philosophy, when held by "one of a reconciling disposition" was not, however, a system: it was an hypothesis (in our sense of that term, and not in the Newtonian sense)[55] which was capable of explaining phenomena, and was in its

[52] Cf. Exp. 38, *ibid.*, III, 250 ff.

[53] This piecemeal and occasional character of his writings may also be seen in another work, "Of the Reconcileableness of Specific Medicines to the Corpuscular Philosophy." There he admits that he is indulging in "a speculative discourse" which only shows "how possibly they [specific remedies] may produce the effects ascribed to them" (*ibid.*, V, 74, 75; cf. p. 108). Since Boyle's purpose in this essay is merely to show that one need not assume that a cure must be effected by the "qualities" of the curative agent, but can flow from its mechanical effects, his disclaimer of any intention of giving in all cases the true causes of the cures should not be interpreted as expressing any skepticism regarding the corpuscular hypothesis itself.

[54] Cf. the "Proemial Essay" of his "Certain Physiological Essays" (*ibid.*, I, 300 ff.) It is no accident that Peter Shaw, who well understood Boyle's thought, should have prefixed this preface to his three volume abridgement of Boyle.

[55] For further discussion of Boyle's use of this term, as contrasted with the Newtonian use, cf. R. S. Westfall: "Unpublished Boyle Papers Relating to Scientific Method," pp. 63 ff. and 103 ff. The discussion of this particular point is to be found on pp. 69–70; cf. also pp. 116–17. [A further discussion of this point is to be found on pp. 103–5, below.]

overall character, the only such hypothesis, according to Boyle. As Thomas S. Kuhn has summarized Boyle's position:

His scepticism and distrust of philosophical system enables him to refuse lengthy dialectic about such metaphysical points as the infinite divisibility of the atom, and the existence of a Cartesian *materia subtilis*; his eclecticism allows him to diverge from both Descartes and Gassendi in developing the corpuscular mechanisms for light, heat, etc.; but neither his eclecticism nor his scepticism extends to doubts that some corpuscular mechanism underlies each inorganic phenomenon he investigates.[56]

---

[56] "Robert Boyle and Structural Chemistry in the Seventeenth Century," p. 19.

There is one passage (*Works*, IV, 232) in which Boyle states that he cannot demonstrate that qualities may not proceed from substantial forms, and that all he wishes to do is to show that they can be explained in terms of mechanical principles. This passage might at first glance be taken as consonant with a positivistic or "sceptical" interpretation of Boyle. However, he then gives reasons which he regards as adequate bases for doubting explanations in terms of substantial forms. Taking this passage as a whole, it seems to me that one is forced to conclude that Boyle is simply cautioning the reader that there is to be no *demonstration*, or (as he also puts it) no *direct proof*, that the doctrine of substantial forms is mistaken; but this is a far cry from any interpretation of Boyle that would hold that the only advantage which the mechanical philosophy possessed over other doctrines is its convenience. To be sure, R. S. Westfall (in the papers cited in the previous note) does interpret this passage as showing that Boyle wavered between accepting the mechanical philosophy as "the one true philosophy" and accepting it as "just a convenient explanation which answers to the phenomena" (pages 107 and 108 respectively), but this interpretation does demand that we assume Boyle to have been contradicting himself in two successive sentences. Instead, my interpretation of this passage would acknowledge that it was neither self-contradictory nor inconsistent with Boyle's usual criticism of substantial forms.

One further point may here be added. In Wiener's article ("The Experimental Philosophy of Robert Boyle," p. 599) a "pragmatic conception of scientific theory" is attributed to Boyle. Clearly, if by such a conception we mean that science is looked upon as useful, and justified in terms of its utility, then Boyle *did* share this conception with Bacon, and with the whole Baconian tradition. However, the usefulness of natural knowledge which Boyle stresses is its usefulness for man's estate. Such a conception of usefulness has not, however, any direct connection with the more modern conception that the meaning of a concept—its "cash value" *within* science—lies in its usefulness in organizing or predicting sensory experiences. (For the most general of Boyle's treatments of the usefulness of knowledge, cf. "Some Considerations touching the Usefulness of Experimental Natural Philosophy," to

To be sure, there is one further passage which might well be cited in reply to my view that Boyle was inexorably committed to the corpuscular philosophy, and that he interpreted it at all times in a realistic manner. This passage immediately follows his remarks on the inability of the ancient atomists to prove that the effects which they noted were caused in the manner that they assumed. In this passage Boyle remarked that "the grand argument" which atomists have used "to confirm the truth of their explications, is, that either the phaenomenon must be explained after the manner by them specified, or else it cannot be explicated intelligibly."[57] To this Boyle says:

in what sense we disallow not, but approve this kind of ratiocination, we may elsewhere tell you. But that, which is in this place more fit to be represented, is, that this way of arguing seems not in our present case so cogent, as they, that are wont to employ it, think it to be . . . . For who has demonstrated to us, that man must be able to explicate all nature's phaenomena . . . . And how will it be proved, that the omniscient God, or that admirable contriver, Nature, can exhibit phaenomena by no ways, but such as are explicable by the dim reason of man?

However, to interpret this passage we must ask what is "this place" to which Boyle refers, i. e., what is the context of this warning that while *in other places* he expects to defend this mode of ratiocination, he will not *in this context* do so? The passage occurs in his treatise on *The Usefulness of Natural Philosophy*, but it occurs in that chapter of the treatise which bears the following heading: "Containing a requisite Digression concerning those, that would exclude the Deity from intermeddling with Matter." In short, what Boyle is here objecting to in the hypothesis of the atomists is the Epicurean view

---

be found in *Works*, II, 1–246 and III, 392–456. Also, cf. *Works*, I, 310–11, where Boyle admits that such knowledge as is derived from particular observations and experiments has been, *up to his time*, more useful than knowledge which could be inferred from speculative systems; it was his hope that in the future "when physiological theories shall be better established, and built upon a more competent number of particulars, the deductions, that may be made from them, may free them from all imputation of barrenness.")

[57] *Ibid.*, II, 46.

of the ultimate origin of the world, and he also objects, at a later point in the same chapter, to the Epicurean view of mind. These are points on which he at all times condemned the ancient atomists,[58] but it is to be noted that he has carefully guarded himself from necessarily condemning their mode of arguing for the corpuscularian philosophy: it is a mode of argument which he promises elsewhere to defend.

I am not aware of any specific passages in which Boyle actually did attempt to prove in detail that only the corpuscular theory was capable of *intelligible* explanations,[59] but a whole series of his works must be viewed as nothing less than attempts to show the superiority of the explanations which the corpuscularian philosophy could give. In the first place, his *Sceptical Chymist* had as its function the destruction of the two major contemporary systems which explained phenomena in terms of irreducible qualities, i. e., the "Spagyrist" and the Aristotelian systems.[60] In addition, there are a great many scattered passages in which explanations by means of substantial forms are criticized,[61] and a special chapter is devoted to this problem in Boyle's most important philosophical work, "The Origin of Forms and Qualities."[62] With respect to the Cartesian system, which was that form of the corpuscular philosophy most remote from his own, Boyle found it inadequate because it did not attempt to explain enough of the qualities in terms of their origins.[63] Thus, he rejected every seriously held alternative to a genuine atomism, and throughout his own work—in instance after instance—he sought to explain

[58] Cf. also *ibid.*, III, 15. Glanvill, one of Boyle's great admirers, makes the same distinction between the atheism of the Epicureans and the new corpuscular philosophy, e. g., in "The Usefulness of Real Philosophy to Religion," *Essays on Several Important Subjects in Philosophy and Religion*, Essay 4, p. 32 f.

[59] To be sure there are passages in which he somewhat dogmatically asserts that if an explanation is not couched in terms of the effects of motion it is not intelligible (cf. below, note 78). However, I assume that the argument which he held to be basic to atomism is more complicated than this. In his own work it assuredly was.

[60] Cf. *Works*, I, 485 ff.

[61] E. g., *ibid.*, p. 411; III, 12 f., 293–97; IV, 232–34.

[62] *Ibid.*, III, 37 ff.

[63] Cf. *ibid.*, p. 11.

a wide variety of qualitative phenomena in terms of the actions of atoms which possessed only those primary qualities which he attributed to them. Among these works may be mentioned his "History of Fluidity and Firmness," his "Experiments and Considerations touching Colours," his "New Experiments and Observations touching Cold," all of whose titles show the qualities with which they are concerned; furthermore, in his "Experiments and Notes about the Mechanical Origin and Production of divers particular Qualities" he took up the mechanical explanation of heat and cold, of tastes, of odors, of volatility and fixedness, of electricity, of magnetism, etc.[64] In none of these explanations of qualities do I find justification for the belief that Boyle merely took atomism as a convenient heuristic hypothesis. I therefore conclude that, for Boyle, the corpuscular philosophy was not merely one way of organizing the relations among phenomena; it was a way of *explaining* these phenomena, where "explanation" is taken to mean tracing these phenomena back to the nonobservable but inferred causes upon which their existence actually depends.[65]

The studies which I have just cited, all of which Boyle refers to as "histories"[66] are attempts to show how the greatest possible diversity

---

[64] The studies here mentioned are to be found in *ibid.*, I, 277 ff., 662 ff.; II, 246 ff.; IV, 230 ff. respectively.

[65] It is important to note that Boyle consistently emphasized the role of inference in explanation, and as Westfall clearly shows in his discussion of Boyle's unpublished papers relating to scientific method, Boyle assigned to reason and not to sense the role of ascertaining the causes of phenomena. ("Unpublished Boyle Papers . . ." pp. 67-68 and 113-15.) (Cf. below, pp. 101 f.)

[66] Boyle wished to follow the Baconian method of giving a descriptive account of all instances of a phenomenon, as Bacon sought to do with respect to heat, so that the comparison of these instances of it could serve as a basis for hypotheses concerning its cause. These accounts Boyle termed "histories" (cf. *Works,* III, 12; IV, 232). It is also to be noted that this usage is similar to that of Bacon in *Novum Organum,* Bk. II, Aphorism xiv (cf. Bacon, *Works,* IV, 145). At one time Boyle intended to write a sequel to Bacon's *Sylva Sylvarum* (cf. Boyle's "Physiological Essays," *Works,* I, 305).

It is to be noted that Locke's use of the phrase "plain, historical method" (*Essay,* Intro., Sec. 2) stands in this tradition. So too does Sydenham's interest in histories. In listing the conditions for the advancement of medicine, Sydenham says, "There must be, in the first place, a history of the disease; in other words, a description that shall be at once graphic and natural"

of qualities can be accounted for by means of his corpuscular philosophy. And in thus accounting for them, Boyle was ridding natural philosophy of the need to appeal to occult qualities.[67] That he was quite self-conscious in his attempt to demonstrate this can be seen in his "History of Particular Qualities"[68] where he considers the objections which might be raised against his doctrine. He finds three such objections, the first of which, based on substantial forms, he had examined elsewhere. Having dismissed it he says: "I enter now upon the consideration of the second, and indeed the grand difficulty objected against the (corpuscularian) doctrine proposed by me about the origin of qualities, viz. that it is incredible that so great a variety of qualities as we actually find to be in bodies should spring from principles so few in number as two, and so simple as matter and local motion."[69] To this he replied by showing that the diversity of sizes, shapes and of motions which are postulated by the corpuscularian philosophy can give rise to innumerable effects, just as the letters of the alphabet, by being suitably arranged, can give rise to all the words of all of the languages of the world.[70] The third objection which he noted is that "if two bodies agree in one quality,

("Medical Observations Concerning the History and Cure of Acute Diseases," in Works, I, 12). Sydenham, however, contrasts histories and hypotheses, and is far more Baconian than Boyle, for he says, "In writing a history of a disease, every philosophical hypothesis whatsoever, that had previously occupied the mind of the author, should lie in abeyance. This being done, the clear and natural phenomena of the disease should be noted—these and these only" (ibid., p. 14). It is clear that in his histories of the various qualities Boyle never sought so radical an extirpation of previously accepted hypotheses from his thought.

[67] Cf. the following statement:

If it be true that . . . the Formes of Divers Bodys be but the result of the determinate size, figure, motion & connection, & suchlike mechanicall Affections of their Component Corpuscles; it will seeme to follow, that since the Occult qualitys of Bodys are resolv'd to flow from their Formes, they likewise may be deduc'd from the same Eminent & obvious Principles; by which if they could be explicated, they would no longer be Occult Qualitys. (Occult Qualitys, Boyle Papers, p. 22; apud Marie Boas, Robert Boyle and Seventeenth Century Chemistry, p. 93.)

[68] Works, III, 292 ff.
[69] Ibid., p. 297.
[70] Ibid., p. 298; cf. Works, IV, 70 f.

and so in the structure on which that quality depends, they ought to agree in other qualities also; since those do likewise depend upon the structure wherein they do agree; and consequently it will scarce be possible to conceive that two such bodies should be endowed with so many differing qualities, as experience shews they may." [71] To this Boyle answered by showing how the corpuscular philosophy assumes great complexity in the bodies which it explains, so that two bodies may agree in that texture which, say, produces the feel of roughness, and yet can disagree with respect to some other qualities; or they may agree with respect to heat, or to whiteness, but disagree with respect to fluidity or solidity. Actually, as he points out,[72] taking into account differences in size, shape, and motion, the corpuscular philosophy admits of far more diversity than does the chemical theory with its three hypostatical principles. In one of the clearest of all of his statements concerning the tenets of his corpuscular philosophy he says:

> We are not to look upon the bodies we are conversant with, as so many lumps of matter, that differ only in bulk and shape, or that act upon one another merely by their own distinct and particular powers; but rather as bodies of peculiar and differing internal textures, as well as external figures: on account of which structures many of them must be considered as a kind of engines, that are both so framed and so placed among other bodies, that sometimes agents, otherwise invalid, may have notable operations upon them; because those operations being furthered by the mechanism of the body wrought on, and the relation which other bodies and physical causes have to it, a great part of the effect is due, not precisely to the external agent, that it is wont to be ascribed to, but in great measure to the action of one part of the body itself (that is wrought on) upon another.[73]

This general statement, appearing in Boyle's "Essay on the Great Effects of even languid and unheeded Local Motion," is followed by attempts to show that from observable phenomena we can prove that "intestine motions" are taking place in solids as well as in fluids; and Boyle shows that there is porousness and motion even in such

[71] *Ibid.*, III, 301.   [72] *Ibid.*, IV, 281.   [73] *Ibid.*, V, 2.

apparently solid bodies as the hardest of woods, in rocks, and in metals.[74] Thus what he attempts to establish is that all qualities of molar matter, and all changes in it, derive from its submicroscopic parts, i. e., from the motions which these variously shaped parts, of varying sizes, undergo.[75]

It is now necessary to ask how Boyle could reach this conviction, or, having reached it, how he could justify it. It is of course usual to assume that this is an epistemological problem, and that, as such, it is not a matter to which the results of scientific inquiry are directly relevant. However, at the time at which Boyle lived the spheres of epistemological and scientific problems were not distinct, and Boyle was able to cite what we would call his scientific results as confirmation for what we would call his epistemological views.

The general tenor of Boyle's epistemology is familiar. As is well known, it was he who first used the terms "primary" and "secondary" qualities in their modern rather than their scholastic meaning. To be sure, Boyle preferred to speak of the affections or modes of body, reserving the term "quality" for our ideas of these affections,[76] but his usage was not consistent in this respect. When using the term "primary" he frequently linked it with "Catholick" (i. e., universal) and with "primitive" and sometimes with "absolute"; in fact, I should say that these terms were synonymous for him.[77] Thus,

[74] "Experiments in the Porosity of Bodies," ibid., IV, 759 ff.

[75] Boyle argued that all intelligible explanation must be couched in terms of motions: motion is the chief cause, shape simply being capable of modifying the effects of motion (ibid., III, 15). As Boyle says, "And if an angel himself should work a real change in the nature of a body, it is scarce conceivable to us men, how he could do it without the assistance of local motion; since, if nothing were displaced, or otherwise moved than before . . . it is hardly conceivable, how it should be in itself other, than just what it was before" (ibid., IV, 73).

This was also Locke's view (cf. Essay, Bk. II, Ch. VIII, Sec. 11). However, due to Stillingfleet's challenge, Locke shifted his emphasis in later editions in order to bring home the point that this might be due to the limitations of our own minds. (Fraser's note to this passage gives the whole of the earlier version of the section, and also gives the gist of Locke's reply to Stillingfleet. For the whole of that reply, cf. Locke, Works, IV, 467–68.)

[76] Cf. Works, III, 292; also, III, 26.

[77] Cf. ibid., pp. 15, 16, 22, 24, 35, 292; also, I, 308; II, 37; IV, 73, 75, 78. For instances of his use of the term "secondary," cf. I, 309 and III, 24.

for example, he speaks of "the primitive and catholick affections of matter, namely bulk, shape and motion.[78] However, all of these terms possessed a temporal as well as an epistemological significance for Boyle, for it was his view that God had originally created "one catholick or universal matter" with its properties of extension, divisibility, and impenetrability, and that the parts of this matter were then set into motion, and out of these universal and primary properties all other differentiating properties arose.[79] The derived properties Boyle conceived of as relations, and he held that such relations do not change the inherent affections of the ultimate parcels of matter. As he said in his famous simile of the manufacture of the first lock and key:

> We may consider then that when Tubal-Cain, or whoever else were the smith that invented locks and keys, had made his first lock . . . that was only a piece of iron contrived into such a shape; and when afterwards he made a key to that lock, that also in itself considered was nothing but a piece of iron of such a determinate figure: but in regard that these two pieces of iron might now be applied to one another after a certain manner, and that there was a congruity betwixt the wards of the lock and those of the key, the lock and the key did each of them now obtain a new capacity; and it became a main part of the notion and description of a lock, that it was capable of being made to lock or unlock by that piece of iron we call a key, and it was looked upon as a peculiar faculty and power in the key, that it was fitted to open and shut the lock; and yet by these new attributes there

---

[78] *Ibid.*, I, 308.

[79] This quasi-cosmological theory is clearly stated in two important passages, one of which is in "The Sceptical Chymist" (*Works*, I, 474) and the other in "The Origin of Forms and Qualities" (*ibid.*, III, 15. Cf. also, III, 35). These passages stand as opening statements of Boyle's general metaphysical principles.

It may also be noted that Boyle, like his contemporaries, did not regard motion as a necessary property of matter, matter being in itself inert. This view was not only characteristic of Boyle and of Descartes, but it was shared by Locke (cf. *Essay*, II, 313) and by Gastrell, among others. (Cf. *The Certainty and Necessity of Religion in General*, pp. 59–61.) Thus Berkeley's argument that matter could not cause ideas because it was inert starts from a commonly accepted principle, although it was used in a way which Boyle and Locke would have rejected.

was not added any real or physical entity either to the lock or to the key, each of them remaining indeed nothing but the same piece of iron, just so shaped, as before.[80]

And so too it is with respect to those qualities which are denominated as secondary qualities: they are produced in us by the action of the primary affections of bodies acting upon our sense organs,[81] but their existence does not modify the bodies which cause them in us.

It must not, however, be believed that Boyle assumed that only the secondary qualities were subjective: so far as perceived size, shape, solidity, and the like were concerned, these too were sensible qualities and were not to be attributed to the objects which caused our ideas of them. As Boyle said: "We must not look upon every distinct body that works upon our senses as a bare lump of matter of that bigness and outward shape that it appears of: many of them having their parts curiously contrived, and most of them perhaps in motion too." [82] This is simply the necessary consequence of Boyle's corpuscular hypothesis, and no other position is consistent with his views on fluidity and firmness, or his views on intestine motions, and the like. The bulk, shape, and motion which constitute the original and primitive properties of bodies, and cause our various ideas of these bodies, are the "affections" of these bodies, and Boyle, as we have noted, distinguished between such affections and our ideas of them.

Now, from the point of view of what we today would call epistemology, one might raise the question of how Boyle could know what constituted the independent affections of bodies, if he took these affections to be different from the ideas which we have when we perceive the various bodies which surround us and upon which we perform our experiments. The answer of course lies in the fact that, according to Boyle, the primary affections of bodies are not sensed,

---

[80] *Works*, III, 18.

[81] For example, *ibid.*, p. 23 f., 31; I, 671, 676.

[82] *Ibid.*, III, 24. Cf. my discussion of Locke's doctrine of primary qualities in "Locke's Realism," pp. 16 ff., above.

but can only be inferred.[83] However, if we then proceed to ask how Boyle could infer the nature of these affections, the way might seem to lie open for an epistemological attack upon his corpuscularianism: it might be claimed that his supposedly transdictive inferences could not be confirmed, or even be understood, unless through reference to what is directly given in sense experience. However, a general epistemological attack of this sort would involve a serious misunderstanding of Boyle's method, for his method—whether rightly or wrongly—did not separate scientific issues from what we should call epistemological issues. Put more concretely, Boyle did not first take for granted a set of scientific descriptions of the world and then ask how these descriptions could be reinterpreted in terms of sensed elements, or ideas, from which they had been derived; rather, his method was to start from his experience with observable objects and to attempt to determine precisely which of their observed characteristics could be used to explain the behavior of objects under varying circumstances, and also to inquire whether there were some characteristics which were not directly observed but which, nonetheless, had to be attributed to objects in order that their behavior should be explained. The upshot of this empirical approach was, as we have seen, a theory of matter, and one which had definite epistemological implications. These implications Boyle accepted as being no less well grounded than the other conclusions concerning material objects which had issued from his experimental inquiries; in fact, he regarded them as following from the latter. In short,

[83] Cf. the following passage from " Of the Positive or Privative Nature of Cold ":

> the organs of sense, considered precisely as such, do only receive impressions from outward objects, but not perceive what is the cause and manner of these impressions, the perception, properly so called, of causes belonging to a superior faculty, whose property it is to judge, whence the alterations made in the sensories do proceed. (*Works*, III, 740.)

However, the " superior faculty " which is here made the judge of the causes of our sensory perception is not to be construed as pure reason operating non-experimentally. In " The Christian Virtuoso," where he is discussing such phenomena as square towers which appear round from a distance, or a stick which appears bent when immersed in water, Boyle explicitly points out that it is not by reason alone, but by *philosophy* (i. e., natural science), that we correct the testimony of our senses (*ibid.*, V, 539).

Boyle's corpuscular theory, for all of its epistemological implications, was regarded by him as being a scientific theory which was grounded upon, and justified through, empirical investigations.

To be sure, as is always the case, some of Boyle's arguments concerning the nature of atoms started from assumptions derived from his predecessors, and were not drawn from his own observations, histories, and experiments; furthermore, some of these assumptions were not themselves based on scientific inquiry.[84] However, his principles of method would not have permitted him to bring forward these very general arguments if he had not also believed that once having made the corpuscular hypothesis we could verify the existence of atoms by means of experiments and other forms of comparative observation.[85] In fact, this is precisely how he proceeded in "The

[84] For example, near the outset of "The Origin and Nature of Forms and Qualities," which was his most careful and most philosophical exposition of the corpuscular philosophy, Boyle lays down three fundamental propositions concerning the material world, and in these he clearly holds that matter exists in its own right, independently of our knowledge of it; that this independently existing matter is the same in all of its exemplars; that its real essence consists in extension and impenetrability; that motion is not part of the essence of matter, but an accident of it; but that this accident (which Boyle, of course, attributed to the direct action of the Deity), is the source of the diversity among what we denominate as individual material objects. (Cf. *Works*, III, 15-16, and the summary statement of the position, p. 35.) In this very general statement of his position there is almost no direct appeal to special observations or experiments. Neither, however, is there any separation of the questions of *whether* matter exists and what qualities it has.

[85] A clear illustration of this is to be found in his "History of Fluidity and Firmness," which we shall cite below in our discussion of his method. In that work he states his main thesis, quotes briefly from Lucretius on the atomist view of fluidity, and then proceeds on his own course of observational and experimental argument.

It would, I think, be profitable to compare the mode of argumentation adopted by Boyle and Walter Charleton's argumentation in *Physiologia Epicuro-Gassendo-Charltoniana*, for there are some striking similarities between them. However, Charleton relies far more heavily upon Greek thought, and upon metaphysical forms of argument. Yet he too seeks to support atomism through empirical inquiry.

I am not at present in a position to estimate the degree of similarity between the empirical arguments; however, it is not necessary to do so in the context of the present discussion, since Charleton's use of arguments which are not directly empirical make the problem of transdiction far less acute for him. With respect to his more empirical arguments, it is my impression that a

Origin of Forms and Qualities": after "the Theoretical part" in which we find his general position stated, and after examining alternative views and special problems, he reaches "the Historical part," that is, the observations and experiments which were designed to verify the corpuscular philosophy. This verification was, of course, highly indirect. Unlike Newton, Boyle did not expect that all knowledge could be directly derived from observation; it was sufficient for him that an hypothesis should be verified through its being *in conformity with* observations. For example, after warning against those who are "over-forward to establish principles and axioms," he adds:

> Not that I at all disallow the use of reasoning upon experiments, or the endeavoring to discern as early as we can the confederations, and differences, and tendencies of things: for such an absolute suspension of the exercise of reasoning were exceeding troublesome, if not impossible. And as in that rule of arithmetic, which is commonly called *regula falsi*, by proceeding upon a conjecturally supposed number, as if it were that which we inquire after, we are wont to come to the knowledge of the true number sought for; so in physiology it is sometimes conducive to the discovery of truth, to permit the understanding to make an hypothesis, in order . . . that by examining how far the phenomena are, or are not, capable of being solved by that hypothesis, the understanding may, even by its own errors, be instructed.[86]

And in another discussion, which was concerned with the confirmation of hypotheses, he said:

> The use of an hypothesis being to render an intelligible account of the causes of effects, or phaenomena proposed, without crossing the laws of nature, or other phaenomena; the more numerous, and the more various the particles [i. e., "particulars"] are, whereof some are explicable by the assigned hypothesis, and some are agreeable to it, or, at least, are not dissonant from it, the more

---

careful study of them would reveal them to be both less systematic and less careful than the analogous arguments in Boyle.

[86] *Works*, I, 303. Contrast Newton's rejection, in the General Scholium, of physical as well as metaphysical hypotheses not deduced from the phenomena (*Principia*, p. 547).

valuable is the hypothesis, and the more likely to be true. For it is much more difficult, to find an hypothesis, that is not true, which will suit with many phaenomena, especially, if they be of various kinds, than but with few. And for this reason, I have set down among the instances belonging to particular qualities, some such experiments and observations, as we are now speaking of, since, although they be not direct proofs of the preferableness of our doctrine, yet they may serve for the confirmation of it.[87]

Taking into account this dictum concerning the confirmation of hypotheses, we may better understand Boyle's desire to amass evidence from the most diverse sources and to show how all of this evidence conformed to the corpuscular view of matter. This desire runs through the largest part of Boyle's scientific writings, and provides a guiding thread for their interpretation.

To be sure, it cannot be claimed that Boyle's *only* purpose in undertaking his various experiments was to establish the corpuscular theory of matter, yet that theory and his particular investigations were almost always closely intertwined. As Marie Boas has shown in her admirable book, *Robert Boyle and Seventeenth Century Chemistry*, such was the case with respect to his chemical experiments, which Boyle himself regarded as being particularly valuable for demonstrating the truth of the corpuscular hypothesis.[88] Now, as we have seen, Boyle was aware of the fact that the evidence for an hypothesis as broad as the corpuscular theory of matter had to be cumulative: no one experiment, or set of experiments, could be decisive in establishing it. Consequently, if we wish to understand how he drew his general conclusion from his experiments and his histories, we must trace out those modes of arguing which repeat themselves in a wide variety of instances, both chemical and physical. To do this in detail would be a difficult task; what I here offer is merely a sketch of some of the most easily noted of the arguments by means of which he sought to link observable phenomena with the corpuscular theory through which he wished to see them explained.

[87] From the Preface to "Experiments, Notes, &c. about the mechanical Origin or Production of divers particular Qualities," *Works*, IV, 234.
[88] Cf. pp. 89–90, *et passim*.

What may be regarded as a first and indispensable step in Boyle's way of connecting his observations with the unobservable hypothetical entities whose actions were to explain phenomena, was to show that our senses are themselves limited in power, and cannot therefore be relied upon to penetrate the secrets of nature. That Boyle should insist upon this limitation is assuredly not surprising, considering the close connection between his thought and that of Bacon. Nor is it surprising when we consider his early interest in the spring and weight of the air, for our unaided senses do not provide us with knowledge of the basic characteristics of the air which constantly surrounds us. Furthermore, in an age in which telescopes and microscopes were relatively new discoveries, one could presumably not fail to be struck by the limitations of the senses. In *Seraphic Love*, an early work, Boyle expatiated upon nature as God's handiwork and spoke of "bold telescopes" through which he surveyed "the old and newly discovered stars," and of how "with excellent microscopes I discern, in otherwise invisible objects, the unlimitable subtilty of nature's workmanship." [89] Furthermore, an interest in chemistry, and in the changes which objects undergo in the course of chemical experiments, is not compatible with taking our unaided senses as an adequate guide to the qualities inherent in bodies. Granted this background, it should occasion no surprise to find Boyle explaining in his "History of Fluidity" why it is that we perceive a fluid such as water in a glass as being one continuous substance whose parts are at rest: the aqueous parts and the pores between them are simply too tiny, and the rapidity of the movement of the parts is too swift, to be discerned by the eye. [90] In the light of this sort of contrast between the grossness of our sense organs and the subtlety of nature, it is to be expected that Boyle would draw precisely the distinction which we have seen that he did draw be-

[89] *Works*, I, 262.

Cf. also I, 676, where he points out that microscopes allow us to discover the roughness of the surfaces of all bodies, and thus lend support to his hypotheses concerning differences in color.

One finds a similar use of the microscope as a support for atomism in Charleton's *Physiologia Epicuro-Gassendo-Charltoniana*, pp. 115, 116.

[90] Cf. *Works*, I, 392.

tween the sensed qualities of bodies and the affections or modes of these bodies themselves, and that he should insist that there was no necessary resemblance between the first and the second. However, this negative doctrine is but the first step in his argument: we must now see by what means he sought to bridge the gap between what we directly perceive and what we can infer on the basis of such observation.

If we follow out Boyle's argument in his "History of Fluidity" we shall note one principle which he frequently uses, and which I shall somewhat arbitrarily term "the extension of sense knowledge by analogy." Obvious and very crude examples of this principle are occasionally found in Boyle's works when he attempts to explain the particular qualities of some type of substance through assuming that its minute corpuscular parts must resemble the characteristics of the sensed whole. For example, he suggests that in order that the corpuscular parts of a fluid may slide and roll easily over one another, these parts must themselves be assumed to be smooth and slippery.[91] However, in general, his use of analogy to extend our sensory knowledge is by no means so crude. For example, in explaining how, on his view, compression accounts for an increase in firmness he uses the analogy of altering the loose texture of new-fallen snow by compressing it into a snowball.[92] In this case the analogy can assuredly carry more weight, for he is not simply asserting that a quality which we perceive a substance to have must also characterize its unperceived parts; he is showing how an operation such as compression, which we know can be performed on an object, and which we can observe as having a particular effect on its macroscopic (or observable) parts, may be assumed to have the same effect on its microscopic parts. However, there are other and more important ways in which Boyle uses the principle of extending our sense knowledge through analogy in his attempts to establish his theory of fluidity. For example, consider the following ways in which he uses analogies drawn from observable phenomena to bridge the gap between what is observable and what is not. First, he cites the fact that under certain unusual conditions of light and of shade we see

[91] Cf. *ibid.*, p. 379.   [92] *Ibid.*, p. 386.

tiny motes floating in the sunlight, or on a hot day there are visible in the air tremulous currents adjacent to walls or spacious buildings.[93] Such observable phenomena suggest that what is ordinarily unobservable may none the less possibly be always present, and that the picture of the corpuscular world of tiny particles of matter in motion has its real analogues in the visible world. Similarly, in this place and elsewhere,[94] Boyle uses the analogy of tiny particles of metals which may be held in suspension in liquids, and which demonstrate the manner in which (as he believes) the unobservable corpuscles of a fluid may themselves be thought to behave.

If it be objected that this use of analogy to extend our sense knowledge is but a fanciful way of proceeding, and lacks any empirical basis, we must take note of the fact that Boyle himself is aware of this challenge, and sets out to answer it. His awareness of it is clear in the following passage:

If it be objected that this various agitation of the insensible parts of water and resembling bodies, wherein we make the nature of fluidity chiefly to consist, is but an imaginary thing, and but precariously asserted, since by our own confession they are so small, that the particles themselves, and more, the diversity of their motions are imperceptible by sense, which represents water, for example, to us as one continued body, whose parts are at perfect rest:

If this, I say, be urged against our doctrine, we shall not deny the objection to be plausible, but must not acknowledge it to be unanswerable.[95]

His answer lies in the fact that he can cite numerous cases in which what we observe concerning the action of a fluid is exactly of the sort which these analogies would lead us to expect. For example, he cites how when sugar or salt is dissolved in a liquid one can taste it in any sample drawn from that liquid, which suggests both an inward and unobserved motion in the liquid and the presence of unobserved particles of salt or sugar throughout it. Similarly, when one puts salt of tartar in a damp cellar its surface will be

---

[93] *Ibid.*, p. 393.   [94] E. g., *ibid.*, p. 380.   [95] *Ibid.*, p. 392.

softened by the imbibed moisture of the air, wherein if it be left long enough, it will be totally dissolved into clear liquor; which would not be, if the moist vapours that help to constitute the air, did not move to and fro every way, and were not thereby brought to the salt, and enabled to insinuate themselves into its pores, and by that means dissolve it, and reduce it with themselves into liquor.[96]

Such observations strongly suggest that the analogies which Boyle has drawn from sense do not lead to a theory which is "an imaginary thing, but precariously asserted."

However, what is perhaps the most striking of the passages in which Boyle tries to bridge the gap between observable molar matter and the unobservably small corpuscles of which he regards material objects as being composed is to be found in his essay "Of the Excellency and Grounds of the Corpuscular or Mechanical Philosophy." There he takes issue with those who are willing to explain effects in mechanical terms when they deal with "Bodies of a sensible bulk, and manifest mechanism," but who none the less evoke "what they call nature, substantial forms, real qualities, and the like unmechanical principles and agents" when they attempt to explain "hidden transactions."[97] Against them he argues:

But this is not necessary; for both the mechanical affections of matter are to be found, and the laws of motion take place, not only in the great masses, and the middle sized lumps, but in the smallest fragments of matter; and a lesser portion of it being as well a body as a greater, must, as necessarily as it, have its determinate bulk and figure: and he, that looks upon sand in a good microscope, will easily perceive, that each minute grain of it has as well its own size and shape, as a rock or mountain. And when we let fall a great stone and a pebble from the top of a high building, we find not, but that the latter as well as the former moves conformably to the laws of acceleration in heavy bodies descending. And the rules of motion are observed, not only in cannon bullets, but in small shot; and the one strikes down a bird

---

[96] *Ibid.*, p. 393.
[97] These quotations and the following passage are to be found in *Works*, IV, 71.

according to the same laws, that the other batters down a wall. And though nature (or rather its divine author) be wont to work with much finer materials, and employ more curious contrivances than art, (whence the structure even of the rarest watch is incomparably inferior to that of a human body;) yet an artist himself, according to the quantity of the matter he employs, the exigency of the design he undertakes, and the bigness and shape of the instruments he makes use of, is able to make pieces of work of the same nature or kind of extremely differing bulk . . . . as a smith, who with a hammer, and other large instruments, can, out of masses of iron, forge great bars or wedges, and make those strong and heavy chains, that were employed to load malefactors, and even to secure streets and gates, may, with lesser instruments, make smaller nails and filings, almost as minute as dust; and may yet, with finer tools, make links of a strange slenderness and lightness, insomuch, that good authors tell us of a chain of divers links, that was fastened to a flea, and could be moved by it; and if I misremember not, I saw something like this, besides other instances, that I beheld with pleasure, of the littleness, that art can give to such pieces of work, as are usually made of a considerable bigness. And therefore to say, that though in natural bodies, whose bulk is manifest and their structure visible, the mechanical principles may be usefully admitted, that are not to be extended to such portions of matter, whose parts and texture are invisible; may perhaps look to some, as if a man should allow, that the laws of mechanism may take place in a town clock, but cannot in a pocket-watch.

Such, I submit, is a clear case in which Boyle uses the principle of extending our sense knowledge through analogy until it is claimed by him that we can grasp the inherent qualities and the forms of action of the insensible parts of which all observable material entities are ultimately composed.

In the above passage it is to be noted that Boyle has not merely used what I have called the principle of extending our sense knowledge through analogy, supposing the corpuscles to have qualities similar to some of those discernible by our senses, he has also assumed that the principles of action which are characteristic of observable entities always obtain among unobservables. This further use of analogical reasoning I shall term "the translation of explanatory principles from the observed to the unobserved." We can find that Boyle makes frequent use of this form of reasoning.

One clear example of how he used this principle is to be found in his account of firmness, in his "History of Fluidity and Firmness." He uses the analogy of two highly polished sheets of glass adhering to one another to explain how the corpuscles in a solid may, if brought into contact, also adhere to one another, and therefore make the object itself a solid.[98] And in general, he assumes that whatever principles explain the phenomena of adherence among observable material objects may also be used to explain the adherence to one another of their corpuscular parts. However, what is probably the most striking case of Boyle's translation of explanatory principles from the observable to the unobservable is to be found in his insistence that all phenomena of the inanimate world are to be explained in mechanical terms, that is, in terms of the transference of motion. As he says in a well-known passage in "The Origin of Forms and Qualities": "I consider that when one inanimate body works upon another, there is nothing really produced by the agent in the patient, save some local motion of its parts or some change of texture consequent upon that motion." [99] While he does not at that point explain in full why he holds this doctrine, his reasons for doing so become plain in his essay "Of the Excellency and Grounds of the Corpuscular or Mechanical Philosophy." [100] Among these reasons there is one which is pertinent here: that only a mechanical explanation is "clear," that is, only it employs principles of explanation which are thoroughly intelligible to us. If we ask why Boyle believed this to be true, we find him holding that in everyday life we only understand how one body can act on another if it acts mechanically, i. e., through transference of motion.[101] Thus, Boyle is holding that in our observation of the inanimate world we find that change depends upon motion, and because he finds that motion and shape can be predicated of any material object, whether it be large or small, he assumes that what holds in this respect in the observable world

[98] *Ibid.*, I, 402–3.
[99] *Ibid.*, III, 25.
[100] *Ibid.*, IV, 67–78.
[101] *Ibid.*, p. 73. For the quotation, cf. note 75, above.

must also hold of the causal relations between the component corpuscles. It is to this assumption that I referred in speaking of the
translation of explanatory principles from the observed to the
unobserved.

However, any such translation can itself be challenged as "an
imaginary thing, and but precariously asserted," unless it receives
some measure of independent confirmation. And it is here that we
encounter what I would call Boyle's third principle, which is, simply,
the method of indirect confirmation which all scientists must employ.
Given a theory—and Boyle *arrived at* his theory by his first two
principles—how is it to be confirmed? The answer, I should suppose, would be that we confirm it by following out its deductive
consequences, and by observing whether these consequences are
corroborated in direct observation and through experiments. This is
precisely what Boyle, in case after case, proceeded to do. Such, for
example, were the results which he obtained in correlating the pressure and volume of gases, upon which the familiarity of his name
now almost exclusively rests. Such too were his observations designed to prove the existence of unobserved, intestine motions in
solids, and his observations with respect to motion as a cause of heat.
To be sure, hypotheses other than the corpuscular hypothesis might
be invoked to explain these phenomena, but it was not the case that
there were any clearly formulated hypotheses which had equal scope
and power, or which seemed to be so thoroughly confirmed. Thus, as
we have said, it is to Boyle that one must look for an attempt to
establish the truth of corpuscularianism, a truth which those who
came after him accepted, and upon which much of their own
scientific work was built.

IV

Enough has been said to show that both Boyle and Newton believed it legitimate to make transdictive assertions. It is now my
intention to give a summary comparison of the grounds on which
each sought to justify that position.

As we have seen, Newton sought to justify his transdictive inferences concerning the qualities inherent in matter by means of Rule III, holding that whatever characteristics were invariably found to be present in experienced material objects may be claimed to be characteristics of all material objects whatsoever. There are two objections which I wish to make concerning this rule. In the first place, let us suppose that in order to explain a particular phenomenon we must make use of what we today would call a scientific "construct." Is it really necessary to hold that such entities possess all of those characteristics which are invariably present in objects which we directly experience? Clearly, our present conceptions of matter do not conform to this rule, and it may even be doubted whether those subtle elastic spirits by means of which Newton himself wished to explain many of the phenomena of nature did possess the hardness which, on the basis of Rule III, Newton attributed to all matter.[102] In short, it is not inconceivable that there might arise a conflict between scientific hypotheses and Rule III.[103] Now, if such conflicts were to arise, it would surely be incautious to suppose that Rule III should be granted precedence over these hypotheses.

In the second place, I should like to point out that Rule III

[102] Cf. the end of the General Scholium appended to the *Principia* (p. 547); also, the following hypothesis concerning static electricity which Newton presented to the Royal Society in his "Second Paper on Light and Colors": Now whence all these irregular motions should spring, I cannot imagine, unless some kind of subtil matter lying condensed in the glass, and rarefied by rubbing, as water is rarefied into vapour by heat, and in that rarefaction diffused through the space round the glass to a great distance, and made to move and circulate variously and accordingly to actuate the papers [i. e., bits of very thin paper with which the experiment on static electricity was carried out] till it return into the glass again, and be recondensed there. And as this condensed matter by rarefaction into an aethereal wind (for by its easy penetrating and circulating through a glass I esteem it aethereal) may cause these odd motions, and by condensing again may cause electrical attraction with its returning to the glass to succeed in the place of what is there continually recondensed; so may the gravitating attraction of the earth be caused by the continual condensation of some other such like aethereal spirit . . . ." (*Isaac Newton's Papers and Letters on Natural Philosophy*, edited by I Bernard Cohen, pp. 180–81).

[103] In this respect Rule III would occupy a different position vis-à-vis empirical explanations than would, say, the principle of the Uniformity of Nature, which presumably cannot be falsified empirically.

places what must be considered to be an unwarranted trust in sense perception. Newton himself was not unaware of the causal chain involved in sense perception, and of the difference between the percept and the nature of the originating factors of that causal chain.[104] Yet, in spite of this awareness he placed his trust in what Rule III refers to as "the senses," or "sensation," or "experience," using sense perception as the foundation for transdiction.

Boyle's method was, I submit, more careful in both of these respects. In the first place, he did not seek to justify transdiction by any special postulate whereby he could move from what is true in all examined cases to what is true in all cases whatsoever. Rather, his transdictive inferences were restricted to attempts to give explanations of the origins of *particular* phenomena in terms of those inferred entities which might reasonably be regarded as their causes. In other words, as we have noted, there was for Boyle no methodological problem of justifying transdiction which was different from the problem of whether a particular scientific inference was a warranted inference. For example, even in the case of his one overarching hypothesis—the corpuscular hypothesis—Boyle accepted this hypothesis, as we have seen, because he believed that it could best explain all of the qualities which he investigated; its confirmation lay in its applicability to case after case. However, we do not find Boyle putting forward any general methodological postulate that the corpuscular hypothesis would therefore have to apply in all cases whatsoever.[105] In short, he believed that transdictions were to be confirmed one by one, by rational inferences drawn from comparative observations and experiments; for him they were part of the *corpus* of science itself, and needed no special methodological justification.

The second point at which Boyle's method of transdiction differed from Newton's lay in the fact that Boyle was unwilling to rest his transdictions on what could be directly confirmed through the senses.

---

[104] Cf. *Opticks*, Queries 12–17, pp. 345–48.

[105] To have insisted that it would have to apply in all cases would have contradicted one of his most fundamental methodological convictions: that system-building was an impediment to experimental inquiries. For his most extensive discussion of this point, cf. the "Proemial Essay" prefixed to "Certain Physiological Essays," *Works*, I, 299 ff.

As we have seen, for Boyle the sensible qualities of objects were simply the effects of the properties of those objects as they acted on our sense organs. Since he did not believe it possible to say with accuracy just why a particular action on one of our sense organs would cause the particular perceptible quality which it did,[106] he did not seek to derive all of our knowledge of these objects from the manner in which they affected us. To be sure, even the secondary qualities which we perceive may *sometimes* give us clues as to changes in properties of the objects themselves, as the changing colors in steel while it is being tempered give evidence of changes in its internal properties.[107] However, the same properties of objects which affect our sense organs also affect other objects, and we can therefore learn at least as much by observing the effects of objects on one another as we can learn by taking note of their sensible qualities, i. e., of their effects upon us.[108] Thus, direct sensory experience did not provide an adequate basis for knowledge of the material world, according to Boyle; as we have seen, it was through our capacity to reason, and to make use of the methods of the new experimental philosophy, that we could extend our knowledge to what lay beyond the limits of sense. To be sure, there were crudities in Boyle's corpuscularianism, and they were in large part due to the first two principles of transdiction which we found him willing to use: the principle of extending sense experience by analogy, and the principle of translating explanatory principles from the observed to the unobserved. However, to those principles Boyle added the classic scientific method of indirect confirmation, and it was upon this method, rather than upon analogies drawn from experience, that he ultimately relied. Unlike Newton, he did not presuppose a single inclusive methodological rule to justify inference to that which is not, and cannot be, directly accessible to sense. Instead, his methods,

---

[106] Cf. *ibid.*, IV, 43–45; also, I, 696. Occasionally, however, Boyle speculated on this problem, as when he suggested that saltiness of taste was directly related to the sharpness and pointedness of the corpuscles (*ibid.*, I, 612; cf. also "Experiments and Observations about the Mechanical Production of Tastes," *ibid.*, IV, 260–61, *passim*).

[107] Cf. *ibid.*, I, 669–70.

[108] Cf. *ibid.*, III, 11, 24. In this Locke follows Boyle: cf. his discussion of powers, *Essay*, Bk. II, Ch. VIII, Sec. 10.

temper, and subject matter demanded that he proceed through individual experimental inquiries, and a comparison of cases, the cumulative effect of which was to establish the corpuscularian natural philosophy as the most inclusive and significant way of accounting for what can be directly observed.

Confidence in the transdictive power of this experimental method can of course be subverted by epistemological arguments, and it has often been thought that Locke's theory of knowledge should have led him to reject all such inferences in spite of the esteem in which he held the achievements of Newton and of Boyle. This, however, is to make the mistake of regarding Locke's analysis of the ultimate origins of our ideas as if that analysis were the whole of his theory of knowledge. While there assuredly are many points at which it does define his position regarding the certainty and the extent of our knowledge, one must also remember that Locke believed that the human mind had great freedom in how it dealt with the elements which it originally derived from experience. In particular one must remember that Locke, unlike Berkeley and Hume, was convinced that human beings possess the capacity to form abstract general ideas. When these matters are not overlooked, Locke's analysis of the ultimate sources of our ideas does not involve a conflict between his views and those of Newton or of Boyle. They too found the ultimate origin of all knowledge concerning the material world to lie in sense experience, and yet they saw the human mind as being capable of inference beyond experience. And he, for his part, agreed with them in holding that science was an autonomous discipline whose credentials were to be found in its achievements. These achievements were linked to the corpuscular theory of matter, and that theory was therefore not to be subverted by argument.

In a succeeding generation, however, the realistic assumptions of seventeenth-century science were challenged on the basis of arguments concerning the principles of human knowledge. In the hands of Berkeley and of Hume the analysis of these principles departed from the more generous views of Locke, and led to a radical reinterpretation of the whole scientific enterprise itself. I shall not be concerned with the historical features of that movement, but in the next chapter shall simply select for discussion some of the arguments which played an important part in it. As we shall see, these particular arguments—even when taken in the subtle form in which they were stated by Hume—do not provide adequate reasons for rejecting some form of epistemological realism. What form such a realism must take, and how closely it in some ways resembles the fundamental convictions of the scientists of the seventeenth century, the final chapter in this volume will attempt to make clear.

3

"OF SCEPTICISM
WITH REGARD
TO THE SENSES"

In The Present Chapter I shall analyze two types of argument which have frequently been regarded as providing adequate refutations of realism. Both were employed by Hume and are to be found in that chapter of his *Treatise* which is entitled "Of Scepticism with regard to the Senses." They were later restated in more succinct form in his *Enquiry concerning Human Understanding*. What I hope to show is that both of these types of argument are invalid when taken as refutations of realism, for at one point or another each presupposes the truth of the very position which it supposedly refutes. After attempting to show that this is the case, I shall discuss two further aspects of Hume's views regarding the possibility of knowing an independent external world. I shall attempt to show that these aspects of his doctrine do not in fact forestall the criticism which I wish to make of his epistemological views.

Before considering the two arguments with which I shall be primarily concerned, it will be useful to define the classificatory terms which I shall use in this discussion, for in some cases they are used in ways that are no longer conventional. I should like to define epistemological *realism* as holding that there exists independently of perception a world of physical objects whose nature can be known by human beings. Direct, or "naïve," realism would hold that the actual qualities of such objects are not different from those which we ascribe to them on the basis of sense perception; in other words, independent physical objects are as they appear to us to be. This position generally involves two connected, but distinct beliefs. First, it involves the belief that objects do possess, independently of our perceiving them, all of the types of qualities which they seem to possess; in other words, no qualities given in sense perception are to be regarded as being, in principle, mind-dependent or mind-engendered. Second, it involves the belief that the specific characteristics which are present to us when we perceive a particular object actually belong to that object. This belief inevitably raises the question of how we are to interpret perceptual errors, illusions, and the like; it may also raise the issue of what role is to be assigned to mental entities when we describe the process of perceiving. While much of the discussion of naïve or direct realism has been centered on these last two issues, the main lines of cleavage between the classic positions in epistemology would seem to demand that emphasis be placed on the question of what, if anything, may be said to exist independently of our perception of it. Therefore, it is upon the first aspects of direct realism that I shall place emphasis in my later discussions of it.

In contradistinction to naïve or direct realism, a critical realist would hold that at least some of the types of qualities which physical objects appear to possess are not actually possessed by them. How radical a distinction is to be drawn between objects as they appear to us and the qualities which are to be attributed to these objects as they exist independently of sense perception, is a matter on which critical realists may disagree. For example, one might hold—as Locke is generally assumed to have held—that in the case of the so-called primary qualities what we perceive not only exists independently of our perception, but exists as we perceive it; that it is only with respect to the so-called secondary qualities that naïve realism is to be rejected. However, a critical realist need not adopt that view. He may hold that the actual qualities of physical objects resemble some of the types of qualities with which we are familiar in sense experience; yet he may also hold that what we are capable of perceiving is never identical with what exists independently of us. This, of course, is the position which I have attributed to Boyle and to Locke, and which I find to be characteristic of seventeenth-century corpuscularianism. However, as we shall see in the following chapter, there may be forms of critical realism which are even more radical than this.

In contrast to all forms of epistemological realism, epistemological *idealism* denies the existence of an independent world of physical or nonmental objects, but affirms that minds or spirits exist and are knowable.[1] However, a more restrained alternative to realism is to be found in *phenomenalism*. Unlike the idealist, the phenomenalist is not concerned to deny the independent existence of physical objects, but only to deny the possibility of any knowledge concerning such objects.[2] Such a denial may take either of two alternative forms. According to one of them, we cannot know *that* there are any objects which exist independently of the contents of our consciousness, whereas the other asserts that such objects do exist, but denies that we can know what their characteristics may be. Since both of these types of position have actually been held, and have been termed phenomenalism, it is perhaps least misleading to define phenomenalism as holding that *if* there are any objects beyond what is contained within our immediate experience we are unable to gain knowledge concerning them.

Now it is obvious that both epistemological idealism and phenomenalism may be further divided in accordance with the arguments which they use, or in accordance with the positions in which they eventuate. However, for the purposes of this essay I shall not be concerned with their variant forms, nor shall I be further concerned with the differences between them. What I wish to do is to con-

[1] By way of caution I should like to point out that I am here only concerned with *epistemological* idealism. That position has sometimes been used to support metaphysical idealism, but the latter can be independent of it.

[2] There are two points to be noted with respect to my characterization of phenomenalism. First, the phenomenalist would also deny the possibility of knowing any substantival mind, soul, or self which lies outside of immediate experience. However, for the purposes of this essay—which concerns only the question of our knowledge of physical objects—that problem may be omitted from consideration. Second, it is to be noted that phenomenalism has recently come to be defined in a quite different way. For example, A. J. Ayer identifies it with "the thesis that physical objects are logical constructions out of sense-data" (*The Problem of Knowledge*, p. 118). There is an obvious connection between the two senses of the term, but what here concerns me is phenomenalism as an ontological position which is comparable to idealism and realism.

centrate attention on a particular thesis which is common to a number of epistemological idealists and to a number of phenomenalists, though by no means to all. This thesis I shall refer to as "subjectivism." [3]

By *subjectivism* I wish to designate the thesis that all we can know on the basis of sense perception are our own "states of mind," or "ideas"—taking the latter term in its broadest, Lockean significance. It is to be noticed that this thesis goes beyond the contention that what we are immediately aware of in sense perception are always our own ideas: that doctrine would be accepted by any realist who held a representative theory of perception. What subjectivism holds is that *all* that we can know on the basis of sense perception are the ideas themselves, the immediate contents of our consciousness; we cannot use them as the basis for inferences to anything which lies beyond them. Now, this thesis will of course lead to an acceptance of phenomenalism unless a claim is made that there also exist ways of knowing which do not rest upon sense perception. Such a claim, of course, characterized the thought of Descartes. While Descartes can be interpreted as having accepted the subjectivist thesis with respect to sense perception, he was able to escape from its usual epistemological consequences by means of an appeal to necessary truths. And Berkeley, too, escaped from those consequences by affirming that there is knowledge not derived from the ideas given in sense perception, namely knowledge through our notions of minds. However, I should suppose that most contemporary philosophers would admit that if one accepts the subjectivist thesis, the most defensible position which one could then hold would be a very cautious phenomenalism. Such a phenomenalism would confine itself to holding that *if* there were anything outside of experience we could not know its nature, and that, in fact, we cannot offer adequate reasons for either affirming or denying that anything of the sort exists. And this, I believe, is precisely the epistemological position which we can best attribute to Hume.

I should, then, be inclined to interpret Hume as holding that

---

[3] Among phenomenalists Kant would not accept what I shall refer to as subjectivism, and Leibniz is an example of an epistemological idealist who attempted to establish that position without reference to the subjectivist thesis.

we must accept the subjectivist thesis, and also as holding that whatever has a right to be characterized as knowledge must be confirmed through appeals to what is given in immediate experience. Consequently, Hume did not escape phenomenalism, nor did he seek to escape it. However, as recent scholarship has led us to see, Hume's position is more complicated than earlier interpretations of it had recognized. For while Hume held that no alternative to phenomenalism can be philosophically defended, he did not identify that which could be philosophically defended with that which is to be accepted in life. In short, as a philosopher Hume was a phenomenalist, but it was only as a philosopher that he accepted phenomenalism: in his other capacities or moods he did not. And this, of course, is merely one example of a fascinating and delicately balanced dualism which is present in many aspects of Hume's thought.[4]

Much as one may be forced to admire his daring and skill, it would, I submit, surely not be a matter of regret if we could avoid having to follow Hume in his adroit maneuvers to escape complete skepticism on the one hand and complete philosophic and scientific

---

[4] Regardless of whether we agree with his interpretation in all respects, the chief person in whose debt we stand for having corrected earlier interpretations of Hume is, of course, Norman Kemp Smith. With respect to the problem of Hume as a phenomenalist, I should like to quote Professor John Passmore, with whose interpretation on this issue I find myself in agreement:

But was phenomenalism the direction in which Hume developed the theory of ideas? Our answer to this question will naturally depend on what we mean by phenomenalism. Laird's definition will serve as a starting-point: "phenomenalism," he says, "is the doctrine that all our knowledge, all our belief, and all our conjectures begin and end with appearances; that we cannot go behind or beyond these; and that we should not try to do so." (Hume's Philosophy of Human Nature, p. 25.) In this sense of the word, I should say, Hume was not, in the end, a phenomenalist, was indeed an anti-phenomenalist; for he regarded phenomenalism as a variety of "excessive scepticism," the sort of scepticism which no one can persistently maintain. We cannot help, whether we like it or not, going beyond "appearances." He was a phenomenalist however, in a narrower sense—he argued that we could not know anything but "perceptions," in that restricted sense of "know" in which it means "be certain of, without any risk of error," nor can we even infer by any sort of "probable reasoning" that anything else exists. So long as he restricts himself to the traditional methods of philosophers, he speaks as a phenomenalist; but this, in his eyes, is part of the evidence that these methods will not suffice. (Hume's Intentions, p. 89 f.)

indifferentism on the other. And this I suggest we can actually do, for I do not believe that Hume's arguments against realism are sound.

I

The first of the two arguments against realism with which I shall be concerned rests on the fact that sense experience frequently offers contradictory testimony concerning the nature of physical objects. The second rests on a causal analysis of the processes involved in perception, and draws a subjectivistic conclusion from the fact that what is directly present to the mind is always and only the last item in this frequently lengthy causal chain. It is not my intention to deny either of the sets of facts to which these two arguments appeal; actually I regard both as true and as important. However, I shall show that these facts are not to be construed as evidence for the conclusion which purportedly follows from them, namely, that sense perception fails to provide us with evidence sufficient to establish knowledge of the existence and nature of an independent physical world. Yet Hume, among others, used these arguments in support of that conclusion.

The most succinct statement of his position on these points is to be found in the *Enquiry Concerning Human Understanding*.[5] There he first says:

> I need not insist upon the more trite topics, employed by the sceptics in all ages, against the evidence of *sense*; such as those which are derived from the imperfection and fallaciousness of our organs, on numberless occasions; the crooked appearance of an oar in water; the various aspects of objects, according to their different distances; the double images which arise from the pressing one eye; with many other appearances of a like nature.

However, Hume does not place his primary reliance on this argu-

---

[5] Cf. Sec. XII, Pt. i. The passages to be quoted are from pp. 151–52 of the Selby-Bigge edition of the *Enquiry*. For the same arguments in the *Treatise*, cf. Bk. I, Pt. IV, Sec. ii and Sec. iv (especially pp. 210 f. and 226 f. of the Selby-Bigge edition of that work).

ment; [6] instead, he then proceeds to the second argument with which I shall be concerned. He introduces that argument by pointing out that men believe, as if by a natural instinct, that the objects which they perceive do have an independent existence; he then continues:

> But this universal and primary opinion of all men is soon destroyed by the slightest philosophy, which teaches us, that nothing can ever be present to the mind but an image or perception, and that the senses are only the inlets, through which these images are conveyed, without being able to produce any immediate intercourse between the mind and the object.

It is with these two arguments, one from contradictions within sensory experience and the other from the causal process involved in perception, that I shall here deal.[7] And I shall deal with them in a way which not only takes into account Hume's statement of them, but which is designed to show what errors are involved whenever they are used to establish subjectivism.

---

[6] He does not do so because he notes that "these sceptical topics, indeed, are only sufficient to prove, that the senses alone are not implicitly to be depended on; but we must correct their evidence . . . ." However, it immediately becomes clear that Hume would reject the possibility of interpreting these corrections in a manner which entails the acceptance of a critical realism. Such a realism, he insists, runs counter to the faith which we repose in the immediate presentations of our senses, and is further weakened by the argument from the nature of the process of perception. Therefore, even though Hume did recognize that the existence of contradictions in sensory perception was not a conclusive argument, he must in my opinion be interpreted as having attached considerable weight to it as a first step in discrediting epistemological realism.

That the foregoing interpretation of this somewhat unclear passage is correct seems likely from the passage in Bk. I, Pt. IV, Sec. iv of the *Treatise*, where Hume traces the origin of "the modern philosophy" (cf. *Treatise*, p. 226).

[7] I shall not here be concerned with the further "sceptical topic" to which Hume attaches weight, namely the argument which he draws from an acceptance of Berkeley's position regarding the impossibility of forming abstract ideas of the primary qualities. (Cf. the last two paragraphs of the first part of Sec. XII, pp. 154–55.) However, as I noted above (p. 116), this constituted one of the major differences between Locke on the one hand and Berkeley and Hume on the other.

A. The Argument from Contradictions

In examining this argument let us first take Hume's own example of the oar which, when partly submerged in water, looks bent but feels straight. It is said that in such a case there is a contradiction. From this contradiction the inference is drawn that our senses sometimes deceive us. Therefore, as Hume points out, if we cannot correct the testimony of our senses by some other means, we cannot know through sensory experience what qualities objects actually possess.

What I here wish to examine is what is involved in the first proposition of this argument; that is, I wish to inquire what it means to say that the fact that an oar looks bent and feels straight involves a contradiction. When I look at the partly submerged oar, I see that at the point at which it enters the water it does seem to form an angle; however, when I run my hand along it there seems to be no angle, and the oar feels straight to my touch. I suppose that all would admit that in those cases in which this actually happens, the pair of observations does involve a contradiction. However, it must be noted that in order that I should interpret these observations as contradictory, there are other beliefs which I must accept. The first of these beliefs is that I am both seeing and touching the same object. The second is that "looking bent" cannot truly indicate a quality inherent in an object if at the same time that object "feels straight"; or, conversely, that "feeling straight" cannot be taken as indicating a quality inherent in an object if at the same time that object "looks bent." In short, at least two sets of beliefs are presupposed if I regard this pair of observations as contradictory: the first is that we can both touch and see the same object, and the second is that certain qualities are incompatible with each other. Each of these points is of sufficient importance to merit discussion, although each may at first glance seem both obvious and trivial.

First, it should be clear that no contradiction exists between seeing a bent oar and feeling a straight one if we are seeing and feeling two different oars. There is only a contradiction if we are touching and seeing one and the same oar. Furthermore, there is only a contradiction if we are, so to speak, touching and seeing the same surface of that oar. This is obvious in the fact that no one would regard it as contradictory if I were to look at the top of a table and find that it looks smooth, and if at the same time I were to touch its under-surface and find that this surface feels rough. Thus, for there to be a contradiction in the experience of seeing an oar which looks bent and yet feels straight, we must actually believe that there is one physical object to which we have access through both sight and touch, and that both of these sense modalities can furnish us with information concerning the same aspect of that object.

Second, in order for there to be a contradiction between our two observations, whatever quality we attribute to the oar on the basis of its looking bent must be incompatible with the existence of whatever quality we attribute to it on the basis of its feeling straight. That there is a contradiction between "looking bent" and "feeling straight" may seem obvious enough: the contradiction seems to follow by definition from what we mean by "bent" and by "straight." However, more than this is actually involved, as can be seen if we examine the theory which Berkeley and Hume would share regarding the relations between our visual and our tactile experiences. Both denied that there is any *necessary* connection between visual and tactile impressions: as Hume frequently reminds us, every distinct perception actually is a separable and distinct existence. The connections which we attribute to these impressions are regarded by both Hume and Berkeley as solely the results of our past experience: it is on the basis of past experience that we have learned to connect certain visual experiences with certain tactile experiences.[8] From this theory it would follow that were it not for past experience, we would find absolutely no contradiction between "looking bent" and "feeling straight." Therefore when we regard it

[8] The classic statement of the problem involved is that of William Molyneux concerning a man born blind who gains the power of sight, and who is then shown a cube and sphere: will he be able to match his visual impressions with his previous tactile experiences, without trial? This problem appears in the second edition of Locke's *Essay*, Bk. II, Ch. IX, Sec. 8. Berkeley uses the case in his *New Theory of Vision*, Sec. 132–33, and draws the same conclusion as had Molyneux and Locke, for their solution of it supports his theory of vision. Hume too would have accepted this solution. Leibniz, on the other hand, rejected it (*New Essays*, Bk. II, Ch. 9).

as "contradictory" that an oar "looks bent" but "feels straight," the contradiction rests on our acceptance of prior sensory experience as showing what qualities of objects are compatible with what other qualities.

The same point may be made more generally—and without relying on what I regard as a dubious theory of space perception and shape perception— if we become a little more specific about the case of the oar in water. In most instances in which this experience occurs, there probably is a time interval between seeing the oar and feeling it; nonetheless, we think that a contradiction is present. And we think it is present because we believe that an oar has certain stable qualities, and that unlike some other objects (such as paper) it does not soften and bend as we gradually submerge it. Furthermore, we may notice that the submerged section of the oar seems to shimmer and waver in ways in which the other part of it does not, and we do not assume that the oar has changed its qualities by virtue of having been submerged. But this, of course, is on the basis of our prior experience with oars, and with objects which resemble oars in other respects. And, finally, if we now slide our hand down the oar, we find that we do not feel that the oar bends at its juncture with the water, but rather that our hand now seems to shimmer and waver, and that our forearm also seems to bend at *its* juncture with the water; yet we do not have any sensations in our hand and arm which seem to be correlated with these changes in visual appearance. Finally, as we gradually withdraw our forearm and our hand, they resume their normal appearance. In all of this, when it is closely examined, there is a great deal which merits discussion, and which suggests that the bent oar is not in fact as happy an illustration of contradictions within sensory experience as might be chosen; however, the point which I here wish to make is a far more limited one. I have only sought to show that in order to claim that an experience of this sort is "contradictory," we must presuppose some knowledge of what qualities of objects of specific types are in fact compatible with each other, and what qualities are not. In other words, if there were not believed to be some order and necessity in the arrangement and sequence of our perceptions, experiences of this sort would not strike us as contradictory, and we would never have any reason to say that our senses deceive us concerning the nature of physical objects.

Thus, there are, I submit, at least two presuppositions which are involved in talking about contradictions in sense experience: first, we must assume that our sensory experience gives us testimony concerning objects which are independent of that experience, and that it is not to be interpreted *merely* as a sequence of ideas in our own minds; second, we must assume that prior sensory experience has given us reliable knowledge of what qualities of objects are compatible with one another, and what qualities cannot be simultaneously possessed by objects of certain types.[9] Were we to reject these assumptions, the so-called contradictions in sense experience would simply disappear.

To be sure, it is not to be inferred from these two preliminary conclusions that Hume and other subjectivists have been refuted: it would be perfectly possible to reinterpret what has here been said in subjectivist terms. What I have attempted to show is that a subjectivist who seeks to *establish* his position on the basis of arguments from contradictions in the testimony of the senses would not in fact have this type of argument to rely upon if, from the outset, he were to interpret the facts which he cites in subjectivistic terms. In order to give one more illustration of this, let us examine those cases in which, as Hume points out, "the various aspects of objects [vary] according to their different distances." Here let us choose two further familiar cases: first, that of a tower which looks round from a distance but turns out to be square; second, the mountains which look blue from afar but become a motley array of colors—green forests and yellowish fields and gray rocks—as one draws nearer to them.

First, we may again note that in order for us to hold that there is a contradiction in our perception of these objects we must assume

---

[9] I do not wish to be understood as maintaining that this knowledge is incorrigible: it is in fact frequently altered in the course of further experience, as one learns when one first plays with "Silly Putty" and finds that what looks like putty can behave in surprising ways. Similarly, one's first experience of an electric shock associates a quite new tactile experience with certain other tactile and visual experiences.

that we are seeing one and the same object when we view it from a distance and when we have approached it. Were we not to assume that it is the same tower which appears first round and then square, and the same mountains which have different colors, no contradiction would be said to exist in our perceptions. This means, however, that we must assume that the tower and the mountains possess a continuing existence. And since it is likely that as we approach such objects we will sometimes lose sight of them, or fail to attend to them, we also must assume that they possess an uninterrupted existence even when we are not perceiving them. Were we to deny these assumptions, there simply would not be a contradiction in the testimony of our senses, any more than there would be a contradiction between looking first at one tower and finding it to be round, and then looking at another and finding that it was square.

Second, we may note that in this case too, beliefs concerning the compatibility and incompatibility of various qualities are presupposed by the judgment that a contradiction exists in the two appearances of the tower, or of the mountain. To be sure, it might be thought that in these cases, unlike the case of the oar, the contradiction involves no assumptions based on prior experience, for " square " and " round," and " blue " and " not blue " would presumably be directly known to be different and incompatible: [10] after all, in these cases, different sense modalities are not involved. However, the matter is not so simple as this. In order to discuss it, we shall have to draw a distinction between two types of cases. The first type would consist of those cases in which the two appearances of the tower, or of the mountain, actually strike an observer as contradictory; the second would be cases in which, though no observer necessarily regards what he sees as being contradictory, epistemologists nevertheless argue from the fact that the appearances are *different*, to the conclusion that our senses may sometimes misinform us as to the actual qualities possessed by external objects.

In those cases in which an observer is actually puzzled, or actually feels that a contradiction exists, the contradiction would not simply be

[10] I have here in mind the doctrine of Locke concerning our ability immediately to discern agreement or disagreement among our ideas, or, as Hume would say, among our perceptions.

a matter of the tower looking first round and then square; in each of his impressions of the tower there would be other discrepant features, for the height would look different, the color might look different, and the amount of detail and the apparent texture of the surface would also appear to be different. In fact, in all likelihood, the only common feature of the "two different towers" would be their spatial location, and it would be on this basis that the "two" towers would in all likelihood be considered not to be two, but one. If the observer were convinced that the tower which he sees when he is nearby is a solid, square tower built of stone, *and* that it occupies the precise place which he formerly thought was occupied by a round tower, then he will say that he was mistaken about the shape in the first instance, for he assumes (and surely not wrongly) that objects of this sort do not shift shape as one approaches them, though mirror images and hedgehogs and many other things frequently do. However, even if he momentarily feels surprise, he is not really likely to remain at all puzzled by the fact that the tower formerly appeared round and is now discerned to be square, since in recalling its earlier appearance he will recall that it not only appeared to be round but that its appearance then lacked the detail which he now sees, that the height of the tower seemed different, and so on. Therefore, the contradictory qualities of the two appearances need not be as puzzling in our ordinary experience as epistemologists sometimes assume them to be,—and as they might indeed be if only one quality underwent a transformation while all of the other qualities and relations of the object remained as they were. But be that as it may, it remains true that there actually are some occasions on which we are surprised by changes in the appearance of something as we come closer to it. In such cases, however, I have already shown that we are only surprised, and we only acknowledge a contradiction in our experience, because of our previous acquaintance with similar objects, and our assumptions concerning the qualities which they possess.

Precisely the same point concerning our assumptions regarding the compatibility and incompatibility of qualities can be made if we now turn our attention to the case of the colors of the mountain. In these cases it is extremely rare (I should think) that an observer actually feels a contradiction between the color which he observes from a distance and the color which he sees when closer at hand. Perhaps this may be due to his past experience in viewing the blue haze that settles over mountains and other distant objects; if so, this constitutes ample support for my thesis that every belief in the existence of a contradiction between two sensory qualities presupposes a belief as to which qualities in objects actually go together, and which do not. And there may be many other reasons why no surprise is in most cases felt in these so-called contradictions. Among these possible reasons I shall cite merely one. We all know how contradictory and bewildering it is when a ball or a handkerchief in the hands of a magician suddenly appears to be a different color from what we expected it to be. In such cases we usually are so convinced that the color of the object cannot have changed that we believe that the ball must be a different ball, that the handkerchief cannot be the same handkerchief. But it is to be noted that in these cases the color of the ball or of the handkerchief is seen as what has been termed " a surface color." [11] However, we are not apt to be bewildered by the change in color which we see as we approach a mountain, nor by what we see as we watch the coloration of a distant mountain changing as the light on it shifts. This is surely in part at least due to the fact that the color of a mountain, when seen from a considerable distance, is more like " a film color " than it is like the surface color of solid objects which we can inspect closer at hand. Whether because of past experience, or for autochthonous reasons, the shifting play of film colors over the surfaces of objects does not lead us to feel that there is any contradiction between seeing an object appearing in one light, or at one distance, as being enshrouded in one color, and then appearing to have another. When, however, conflicts do actually arise with respect to the perceived colors of objects, as they do in the case of the magician's tricks, it becomes all the more evident that we only hold two appearances to be contradictory

[11] On the phenomenology of colors, cf. David Katz, *The World of Colours* (London, 1935), a translation of the second edition of *Die Erscheinungsweisen der Farben*, which originally appeared in 1911. It is important to note that the term " surface color " is here used in an entirely different sense than it has in physical optics.

on the basis of beliefs as to which sensory qualities are to be regarded as the stable and continuing qualities of the object, and which sensory qualities are merely evanescent appearances compatible with the stability of other qualities which the object possesses.

Though this point might be admitted with respect to the existence of felt contradictions on the part of observers in real-life situations, an epistemologist might nonetheless argue that once we admit that the same tower can look round from a distance and square when seen close at hand, or that the same mountain can look now one color and then another, we are no longer able to maintain that the testimony of the senses is a reliable guide to the nature of objects. His argument would be that if the senses can sometimes deceive us by virtue of giving us differing reports, it is at least theoretically possible that they always do so; or, at the least, he can challenge us to produce any clear criterion by means of which we can in every case know when our senses deceive us, and when they do not.

However, neither form of this argument can escape the objections which I have been raising. What I have been attempting to show is that one cannot prove that the senses actually *do* sometimes deceive us without assuming that they sometimes do not. I would therefore contend that this skeptical argument is self-refuting. It consists in drawing the conclusion that we can never know whether our senses are deceiving us from the fact that sometimes they actually do deceive us; however, as I have argued, this premise—the statement that they do sometimes deceive us—could not itself be known to be true if the conclusion of the argument, that we can never know whether they are deceiving us, were itself taken as true.

Against this answer it might be contended that I have been assuming throughout this discussion that it is only through further sensory experience that we can know that our senses have deceived us, and it might be held that this claim is false. Some would claim it to be false because they would claim that once it is granted that a tower may appear as being either round or square, the fact that these predicates are contradictory leads one immediately to the conclusion that whatever organ is capable of leading to mutually contradictory conclusions is not to be trusted as an organ of knowledge. However, there is an error in this rejoinder, and it consists in a confusion between what is logically contradictory and what we mean when we speak of the contradictions inherent in sense experience. This can be shown in the following way. " Blue " and " not blue " are logically contradictory, and when we understand the meanings of " straight " and " bent," or of " round " and " square," we can also see that we cannot without contradiction attribute both of these predicates to the same object in the same sense at the same time. No one, however, has ever claimed that his senses directly presented him with instances of logical contradictions of this sort. The contradictions arise when we perceive what we take to be the same object, or the same region of space, through two different sense modalities, or at different times, or under different conditions. To classify our sense experiences as involving contradictions something more than the merely logically contradictory nature of the two qualities is therefore needed: [12] we must also make assumptions concerning the nature of physical objects, of the sense organs, of the compatibility and incompatibility of different qualities under different conditions, and the like. Therefore, the epistemological argument which we have here examined is not in fact independent of the argument from the judgments of actual observers which I previously examined.

It will be recalled that, in the case of the oar, I argued that even to speak of a contradiction in sensory experience involves two fundamental assumptions: (1) that we interpret our senses as giving testimony for the existence of objects which are independent of our experiencing them, and (2) that we assume that our prior sensory experience has given us reliable knowledge concerning some of the qualities of these objects. It is for this reason that I regard an argu-

---

[12] It is also to be noted that not all contradictions in the testimony of the senses involve attributing logically contradictory qualities to the same object or region of space. For example, we know through the use of X-ray photography that entities can penetrate our bodies without our feeling them, and without causing any visible changes on the surface of our bodies; yet, if we saw a knife apparently pass through a person's finger and we saw no cut, and he exhibited no sign of pain, we would (with warrant) think that our sense of sight had deceived us. (And such phenomena can be generated through illusions dependent upon stroboscopic movement.) In cases of this sort, the contradictory testimony of our senses does not involve the question of qualities which are logically incompatible with one another, but only those which are empirically so.

ment for subjectivism which is based on the existence of contradictory sense experiences as a self-refuting argument: our very belief that there are genuine contradictions in sensory experience would be an unwarranted belief if subjectivism were true. And I have now attempted to show that this is the case not only with respect to the phenomenon of the submerged oar which looks bent but feels straight, but also with respect to contradictions which involve variations in the appearances of objects when seen from varying distances. Further, it is to be noted with respect to these two sets of illustrations, both of which are widely cited by others as well as by Hume, that the same conclusion was reached whether one or more sense modalities were involved, and whether the qualities being discussed were so-called primary qualities of shape or were differences in color. I therefore think it safe to conclude that my argument can be generalized to fit any case in which the contradictory testimony of the senses is cited, and safe to say that such cases offer no basis for the conclusion that subjectivism is true.

However, lest I be misinterpreted, let me state—and it is a point to which I shall later return—that the insufficiency of this argument in favor of subjectivism does not prove that subjectivism is false. And let me also remind the reader that—as we have already noted—Hume himself did not place his primary reliance upon it. Rather, he relied more heavily upon the implications which supposedly follow from a causal analysis of the perceptual process.[18] It is to a consideration of this argument that we must now turn.

B. *The Argument from a Causal Analysis of Perception*

One of the best known of Hume's statements concerning his acceptance of subjectivism, and perhaps his most rhetorical formulation of that thesis, is to be found in his discussion " Of the Ideas of Existence, and of External Existence " in the *Treatise*. There he says:

---

[18] In *Human Knowledge, Its Scope and Limits*, Bertrand Russell discusses his own acceptance of Hume's thesis that " all my data are private to me," and says that in his acceptance of that thesis he " attaches special weight to the argument from the physical causation of sensations " (p. 174).

Now since nothing is ever present to the mind but perceptions, and since all ideas are derived from something antecedently present to the mind; it follows, that it is impossible for us so much as to conceive or form an idea of anything specifically different from ideas and impressions. Let us fix our attention out of ourselves as much as possible; let us chase our imagination to the heavens, or to the utmost limits of the universe; we never really advance a step beyond ourselves, nor can conceive any kind of existence, but those perceptions, which have appeared in that narrow compass. (pp. 71–72)

Hume sometimes speaks as if we directly knew that this was the case, and that no argument were needed to establish it: for example, he says " Since all impressions are internal and perishing existences, *and appear as such . . .*" [14] This, however, is surely a mistake, and runs counter to what he himself holds concerning our common-sense realism.[15] The doctrine that all of our impressions are internal and perishing existences, existing only in the mind, is not something of which we are directly aware when we see a chair or a tree; it is the product of reflection or analysis. And this, I submit, is Hume's own doctrine. What he holds is that a knowledge of the subjectivity of all experience is the result of " philosophy." For example, he says: " We may observe, that it is universally allowed by philosophers, and is besides pretty obvious of itself, that nothing is ever really present

---

[14] *Treatise*, Bk. I, Pt. IV, Sec. ii (p. 194); the italics are mine. Cf. also " Add to this, that every impression, external and internal, passions, affections, sensations, pains and pleasures, are originally on the same footing: and that *whatever other differences we may observe among them, they appear, all of them, in their true colours, as impressions or perceptions*" (p. 190; the italics are again mine).

[15] For example, in the same general discussion (*ibid.*, p. 192) he distinguishes " three different kinds of impressions conveyed by the senses ": the first are those of the so-called primary qualities, the second those of the secondary qualities, and the third those of pleasures and pains. Of these he says " Both philosophers and the vulgar suppose the first of these to have a distinct continued existence. The vulgar only regard the second as on the same footing. Both philosophers and the vulgar, again, esteem the third to be merely perceptions; and consequently interrupted and dependent beings." This obviously contradicts the statement that all of our impressions actually appear to us as " internal and perishing existences " and that all of them appear to us as being, originally, " on the same footing."

with the mind but its perceptions or impressions and ideas, and that external objects become known to us only by those perceptions they occasion." [16] And again: "Philosophy informs us that everything which appears to the mind is nothing but a perception, and is interrupted and dependent on the mind." [17] And, as we have noted, he speaks in a similar vein in the *Enquiry*, where he remarks of our common-sense realism that "this universal and primary opinion of all men is soon destroyed by the slightest philosophy, which teaches us, that nothing can ever be present to the mind but an image or perception, and that the senses are only the inlets, through which these images are conveyed."

It is not entirely clear what Hume has in mind when he speaks here of "philosophy"; however, I believe that one can best interpret him as intending to refer to a causal analysis of what occurs when we look at, listen to, touch, taste, or smell objects. That this is a reasonable interpretation is attested by both the first and the last of the three passages which I have quoted, for in them Hume clearly has in mind the fact that our perceptions are but the last step in a causal chain. To be sure, Hume holds that the detailed analysis of this process belongs to "anatomists and natural philosophers"; nonetheless, scientific inquiries of this sort were not so remote from philosophic problems for Hume and for his contemporaries as they have since become. [18] The account which Hume presents of our

---

[16] *Treatise*, Bk. I, Pt. II, Sec. vi (p. 67).

[17] *Ibid.*, Pt. IV, Sec. ii (p. 193).

[18] The breadth of Hume's use of the term "philosophy" may be traced through the indices of Selby-Bigge's editions of the *Treatise* and the *Enquiries*, but special reference should also be made to Hume's introduction to the *Treatise* which is not indexed by Selby-Bigge. It is clear that Hume's usage is far broader than that which includes "natural philosophy" only; on the other hand, natural philosophy (or, as we should often say, "science") is included as one part of philosophy. Hume's usage of the term "philosophy" may in my opinion be equated with Berkeley's. The latter opened the introduction to his *Principles of Human Knowledge* by contrasting those who pursue philosophy with "the illiterate bulk of mankind that walk the high-road of plain common sense" Philosophers are in effect identified by Berkeley with all those who "depart from sense and instinct to follow the light of a superior principle, to reason, meditate, and reflect on the nature of things," and philosophy then can be claimed to be an analytic and reasoning approach to what Berkeley called "the study of wisdom and truth." For

senses as being merely "the inlets" through which images are conveyed to the mind, and his further contention that what we perceive are merely these images, thus conveyed, and not the objects themselves, actually is of a piece with the scientific account of the causal chain in perception; it is certainly not what we would ordinarily believe on the basis of direct experience alone. This may perhaps best be documented by considering for a moment what it is that direct experience seems to suggest concerning the role of the senses in the perception of objects.

Without consulting "anatomists and natural philosophers" I can surely be said to learn through a comparison of cases that what I perceive depends upon the activities of my sense organs. For example, I presumably know that when I close my eyes I can no longer see the tree at which I have been looking, and when my eyes are open I cannot voluntarily efface the sight of a tree which is immediately in front of me. Similarly, if I stop up my ears, the sound which I have been hearing becomes muted, yet I cannot "think away" a sound: I must shut it off by taking steps to preclude it from reaching my ears. In all of this our ordinary experience suggests that our eyes and ears and other peripheral sense organs are the means by which we perceive that which we do perceive. But this fact is perfectly compatible with common-sense realism. In ordinary experience we regard our senses as the channels through which we receive impressions of the world, but we nonetheless believe that what we are perceiving is external to us and independent of our perception of it. In other words, we are apt to think of the senses as windows or doors by means of which we have access to what exists outside of ourselves. There are, however, many reasons why this belief cannot be maintained. Among these reasons is the fact that scientific inquiry shows that what is transmitted from the bell to my ears is not "sound" but certain waves in the air, and that the color which appears to be located in objects is also transmitted in a form quite different from the manner in which it appears. Similarly, when we consider what must occur within our own bodies for us to be cog-

Hume, too, philosophy seemed to embrace all reflective analytic inquiry, whether applied to the findings of experimental inquiries, to common experience, or to accepted beliefs.

nizant of what is affecting our sense organs—how nerves must not be severed, and how (to use seventeenth-century terminology) " animal spirits " must flow to the brain—we recognize that our contact with external objects is not so direct and open and altogether obvious as the simile of windows and doors would lead us to expect. What natural philosophy and anatomy therefore show is that our sense organs play a different and more complicated role in the process of perception than we directly experience them as playing. In order that we should perceive external objects these organs must themselves be affected, and they must then transmit physical effects which are quite different from the things of which we are immediately conscious.[19] This dual role of the sense organs in perception was understood by Locke who, in stating his doctrine of the origin of knowledge in sensation, said: " Since there appear not to be any ideas in the mind before the senses conveyed any in, I conceive that ideas in the understanding are coeval with *sensation; which is such an impression or motion made in some part of the body, as produces some perception in the understanding.*"[20] To fill out the details of how the sense organs operate would, of course, be to engage in " anatomy and natural philosophy," which neither Locke nor Hume wished to do. Their reluctance on this score cannot, however, be attributed to any distrust of the general account of these matters which had been given by Descartes and by other modern philosophers; rather, they trusted such accounts sufficiently not to have troubled to defend them. Consequently, even " the slightest philosophy "—or, as I should interpret this phrase, the merest acquaintance with these

---

[19] It is "animal spirits" and not shapes and colors which are transmitted through the nerves, but it is shapes and colors and not the animal spirits which we see.

[20] *Essay*, Bk. II, Ch. I, Sec. 23. The italics are Locke's own, but the italicized portion of the sentence gives Locke's view of the dual role of the sense organs in perception, since he is here equating " sensations " with what occurs on the sense organs. (The term " produces " was substituted in the third edition for "makes it to be taken notice of," but the meaning is the same.) In at least one incautious passage Hume shows how near his own views were to Locke's, for he says: " The most vulgar philosophy informs us, that no external object can make itself known to the mind immediately, and without the interposition of an image or perception " (*Treatise*, Bk. I, Pt. IV, Sec. v, p. 239).

results—was sufficient to prove the complicated nature of the causal chain which is involved in sense perception. However, it is to be noted that it does take inquiry—that is, it does take philosophy—to establish such a causal chain: direct experience frequently seems to suggest the contrary notion, that our senses are like open windows and doors through which the outside world makes direct contact with our minds. In my opinion, it was this causal account of the process of perception that Hume had primarily in view when he spoke of " philosophy " as having established that what we perceive is actually not an independently existing object, but is in all cases an internal and perishing existent.[21]

I do not believe that the fact that a causal chain is always involved in sensory perception can legitimately be used as the basis for an argument in support of the subjectivist position; on the contrary, I believe that such arguments, though frequently so used, are fallacious. This is not because a subjectivist cannot interpret the causal chain in perception in terms of subjectivism, once he has accepted that position, for he can. Therefore, the causal argument is not self-refuting in the sense that we have found the argument from contradictions to be self-refuting: in the latter argument one could not consistently hold that all that we perceive are only ideas in our own minds and also hold that there are actual contradictions between what we perceive at different times. In the case of the causal analysis of the perceptual process this inconsistency does not exist. However, I shall show that there would not actually be

---

[21] In this connection it is to be recalled that Hume did not believe that the argument from contradictions was sufficient to prove this thesis. His use of Berkeley's attack on abstract ideas to reject the possibility of holding that the primary qualities are objective, even though the secondary qualities are not, is of course a further argument for subjectivism; nonetheless, it depends (in part at least) on an acceptance of the causal account of sense perception in order to establish the subjectivity of the secondary qualities (cf. his analysis of the contradictions among the secondary qualities, *Treatise*, Bk. I, Pt. IV, Sec. iv, pp. 226–27). What is identified as " the modern philosophy " by Hume involves the causal analysis of perception, and Hume always accepts that philosophy as a first step in the direction of truth, but then goes beyond it. It is for these reasons that I am inclined to interpret Hume as holding that the crucial arguments for subjectivism—the argument from " philosophy "— rest upon a causal analysis of the perceptual process.

evidence adequate to establish the existence of the causal chain in perception unless one were willing to accept some form of the realistic hypothesis. In other words, I shall be concerned to prove that any well-grounded belief that there is the sort of causal process in perception which "the least philosophy" shows that there is, presupposes that we regard the links in that chain as events which are not mind-dependent. If this is so, it is fallacious to hold that the causal analysis of perception provides a good argument in favor of subjectivism.

As I have said, Hume was not himself concerned to establish the precise nature of the causal processes which are involved in perception, but he apparently did not question the sort of analysis which Descartes and then Malebranche offered with respect to the series of bodily events which were involved. However, there was one point at which he clearly did have in mind a specific theory concerning one of the aspects of sense perception, and this was his acceptance of Berkeley's theory of vision. He used this theory as a means of undercutting the obvious and insistent common-sense objection to subjectivism which is based on the fact that we see objects as external to us, that they are "out there" and not in the mind. After taking note of this argument, Hume replied: "Our sight informs us not of distance or outness (so to speak) immediately and without a certain reasoning and experience, as is acknowledg'd by the most rational philosophers." [22]

---

[22] *Treatise*, Bk. I, Pt. IV, Sec. ii (p. 191). The same position is implicit throughout his treatment of our ideas of space in Part II of Book I of the *Treatise*, and not merely in his treatment of our ideas of external existence at the end of that Part. And in one passage in Section v of his discussion of space he says: "'Tis commonly allowed by philosophers, that all bodies, which discover themselves to the eye, appear as if painted on a plain surface, and that their different degrees of remoteness from ourselves are discover'd more by reason than by the senses" (p. 56). Cf. also his two additional comments concerning the perception of distance in the Appendix to the *Treatise*, pp. 632, 636.

It is furthermore to be noted in support of my contention concerning the connection between those whom Hume designates as "philosophers" and those who are concerned with "anatomy and natural philosophy" that in both of the above passages which I have quoted, "philosophers" must obviously know optical theory and also the anatomy of the eye.

The theory to which Hume here had reference was clearly that of Berkeley, who held that our perception of distance (i. e., of the third dimension) must be the result of judgments based on past experience, for distance could not, according to him, be directly perceived. The reason why he held that it could not be directly perceived is explicitly stated in the second paragraph of his *Essay Towards a New Theory of Vision*. There he says: "It is, I think, agreed by all that distance, of itself and immediately, cannot be seen. For distance being a line directed endwise to the eye, it projects only one point in the fund of the eye, which point remains invariably the same, whether the distance be longer or shorter." If the third dimension is not represented on the retina, how then do we come to see objects as placed at a distance from us? Berkeley regards it as well established that in the case of remote objects our judgment of the distance depends upon such perceptual cues as the number of intervening objects, the apparent smallness and faintness of the objects at which we are looking, and the like; our interpretation of such cues being of course based on past experience. However, Berkeley points out that those who are willing to accept this analysis with respect to remote objects are not willing to do so with respect to nearer objects; instead, they hold that in the latter cases our judgment of the third dimension depends upon the angle of the two optic axes, that is, upon the binocular parallax.[23] It is against this hypothesis that Berkeley argues. He argues against it on the grounds that since we are not aware of these angles, we could not use them as the basis for a judgment as to the distance at which an object is placed. Consequently Berkeley claims that the whole of our knowledge of the third dimension is founded on the orderliness of our past experience with respect to

[23] As A. A. Luce points out in his edition of the *Essay* (*The Works of George Berkeley, Bishop of Cloyne*, I, 171, n. 2), Berkeley is much indebted to Molyneux, who takes precisely the view here stated. In fact, the passage quoted from Berkeley concerning distance being "a line directed end-wise to the eye" is a paraphrase of Molyneux (cf. *Dioptrica Nova*, Prop. XXVIII, p. 113). Berkeley's reliance on Molyneux makes the argument which I shall offer against Berkeley's subjectivism all the more cogent, for Molyneux's treatment of vision obviously presupposes the acceptance of epistemological realism: the objects which we see are never considered by him as analyzable into sets of ideas in our minds.

the relations obtaining between "genuinely" visible qualities such as color, size, shape, visible detail, etc., and our kinaesthetic sensations, plus our sensations of touch. On the basis of his belief that this is the true explanation of our ideas of distance, Berkeley makes the following uncompromising statement with respect to the problem which Molyneux originally propounded to Locke:

> It is a manifest consequence that a man born blind, being made to see, would at first have no idea of distance by sight; the sun and stars, the remotest objects as well as the nearer, would all seem to be in his eye, or rather in his mind. The objects intermitted by sight would seem to him (as in truth they are) no other than a new set of thoughts or sensations, each whereof is as near to him as the perceptions of pain or pleasure, or the most inward passions of his soul.[24]

I am not here concerned with the truth or falsity of the Berkeley-Hume theory of our perception of distance,[25] but only with examining the grounds on which this theory was put forward. These grounds were explicit in the passage which I have already quoted from the opening of the *Essay Towards a New Theory of Vision*. In that passage Berkeley spoke of distance as being "a line directed end-wise to the eye," and he spoke of such a line as being projected only as "one point in the fund of the eye" (i. e., as one point on the retina). The first of these statements presupposes a knowledge on Berkeley's part of the existence of rays of light reflected from the surfaces of objects, and of the propagation of these rays in a straight line; the second presupposes a knowledge of their mode of refraction

---

[24] *Essay Towards a New Theory of Vision*, Sec. 41.

[25] However, I do not believe that it can any longer be regarded as possibly true. A classic experiment which was designed to test whether the visual perception of distance was acquired by learning seems firmly to establish that it is not: cf., Lashley and Russell, "The Mechanism of Vision. XI. A Preliminary Test of Innate Organization." For a summary of recent findings concerning the factors responsible for stereoscopic vision, cf. Kenneth N. Ogle, "Theory of Stereoscopic Vision," in *Psychology: A Study of a Science*, I, 362–94. For a solution of the problem as to why our percepts appear as localized outside of our bodies, a problem which is closely related to the manner in which Hume utilized Berkeley's theory of vision, cf. W. Köhler, *The Place of Value in a World of Facts*, pp. 127–41.

in the eye and the manner in which they are focussed on the retina. Let us now see what an acceptance of these propositions presupposes when they are taken as true and reliable analyses of what occurs in vision.

We may first note that in any case in which we are directly perceiving remote objects, we are not at the same time conscious of light rays, their properties, and the manner in which they are being focussed on our retinas. We must also note that in any case in which an ophthalmologist inspects our eyes *he* is not perceiving the objects which *we* see in the remote distance: through his ophthalmoscope he is merely seeing their inverted images on our retina. Consequently, there is no directly experienced relation between the two sets of observations which Berkeley wishes to correlate: that is, between those observations of distance which we think are directly given to us, and those observations which are made by physicists who are interested in light and physiologists who are interested in the physiology of vision. To be sure, the two sets of observations can almost always [26] be correlated: when we see something as outside of us, the physicist can find that light rays are reaching our eyes, and an ophthalmologist can find imaged on our retinas the projections of what we describe ourselves as seeing. However, our belief that such a correlation is *always* present would have very little inductive support if it rested only on the number of cases in which it has been observed to occur. The slightness of this evidence can be noted simply by asking how many times we have the experiences requisite to correlating the perception of distance with the facts cited by Berkeley, compared to the total number of cases in which we perceive distance without having evidence of the existence of light rays, their refraction in the eye, and the manner in which they are projected on the retina. Therefore, the anatomical and physical facts which Berkeley and Hume take for granted in their theory of vision can only be regarded as reliable if we are willing to assume that the objects and processes which are involved are not mind-dependent,

[26] It would not be strictly accurate to say "always," but I shall disregard the exceptions. The fact that there are exceptions would fortify my point, rather than weakening it.

but are present independently of perception and occur even when they are not observed as occurring.[27]

A second way in which we may show that this causal account of vision presupposes an acceptance of realism is to be found in the difference between the way in which it treats our sense organs and the way in which we directly experience these organs as operating in everyday life. As has already been noted, direct experience seems to show that our sense organs are channels through which we have contact with the external world. For example, I find that I cannot "think away" a tree which appears directly before me, although I can stop seeing it by merely closing my eyes. However, were we to reinterpret our sense organs as themselves being only sets of perceptions which are mind-dependent, all that we could say that we actually know of a causal process in perception is that when we do see a tree we also have the experience which we call having our eyes open, and when we have the experience of shutting our eyes we do not have the experience of seeing the tree. However, the causal account of what occurs in vision involves far more than this correlation of sets of perceptions. It involves holding that light is reflected from the surfaces of objects, and that when our eyes are open this light penetrates the lenses of our eyes, and is focussed on our retinas, and affects them in ways that lead us to have visual impressions of objects. All this is not something of which we have direct experience. Thus, if we interpret our sense organs as themselves being only sets of perceptions which have an internal and perishing existence in our own minds, we would not actually have good grounds for accepting those physical and physiological beliefs concerning vision which Berkeley and Hume took for granted as applying in all cases to all persons, and upon which the Berkeleian theory of the perception of distance was based.

Once again I must insist that the foregoing argument is not intended to prove that Hume would be unable to interpret all of the evidence of physics and of physiology in subjectivistic terms. Not only might he do so, but he could presumably explain in terms of

---

[27] The same point is made, although in a slightly different way, by H. H. Price in *Hume's Theory of the External World*, pp. 115–16.

his theory of belief just why we believe that light waves and retinas exist when we are not perceiving them.[28] In other words, if we were simply to grant Hume the truth of subjectivism and of his theory of belief, there would be no inconsistency in his also holding the theory of vision which he did hold. However, the point which I wish to make is a different one. What I have wished to show is that if one *first* accepted subjectivism, holding that our sense organs themselves were mind-dependent, there would be insufficient evidence in favor of the hypotheses which Hume adopts concerning the conditions under which distance is perceived. Consequently, I would claim that it is illegitimate to use this theory of vision as an argument in favor of subjectivism, as both Berkeley and Hume were inclined to do. Furthermore, the same point can be made more generally, in ways applicable to any form of sense perception. As I shall now show, the contention that what is directly present to us in sense perception is only the last link in a causal chain would not be a plausible contention unless each of the links in that chain were itself regarded in a manner consonant with realism. That this is the case can be seen from a general analysis of the types of links which exist in the causal chain in perception.

In sense perception the usual types of physical factors involved are (1) a specifically located object which appears to be outside of us, (2) a medium between us and that object, (3) our peripheral sense organs through which the object affects us, (4) the nerves leading from the peripheral sense organ to a particular area of the brain, (5) some processes in the brain which we may term "perceptual processes," since they are correlated with the act of perceiving. To be sure, in some cases the first of these five factors may not be specifically localized, and in other cases, such as touch, there

---

[28] I say "presumably," but it is to be noted that Hume never gives a concrete analysis of why we believe in the independent and continued existence of entities such as light rays or retinas or "animal spirits": his analysis of our belief in bodies is confined to cases such as our belief in the independent and continued existence of objects such as stairs, and letters, and people. It is presumably possible that "constancy" and "coherence" may give rise to our belief in the presence of light rays in all cases in which vision occurs, etc., but anyone seriously interested in maintaining Hume's account of the matter should surely examine such questions in more detail.

may not be a medium between us and the sensed object. However, we may take these five factors as outlining the usual types of links which exist in the causal chain which is present when we perceive an external object. In perceiving parts of our own bodies (for example, when I see my own hand) all of the same factors may also be present, although in other cases (for example, in kinaesthetic sensations or in experiencing a localized pain) the chain may lack its earlier links. Without the last link, however, perception does not occur. It is for this reason that the causal chain which is involved in perception has so often been interpreted as an argument in favor of subjectivism. What we directly experience, have contact with, or know (which word we use is not in itself crucial), must be regarded as being caused by perceptual processes in the brain. Since this last link occurs "within us," it would seem that what we know is not an independent world, but is always and only states of our own consciousness. Thus the causal chain in perception appears to support subjectivism.

Yet it only appears to do so. It is not by direct experience but only by scientific inference that we know that such a chain exists. And if we could not trust such inferences we would have no adequate grounds for believing that what we directly experience, or have contact with, or know, is only an idea in our minds. These inferences, however, are not merely taken as referring to what occurs within our own direct experience: they are taken by us as referring to something which transpires outside of the content of our consciousness. For example, knowing that brain activity is being stimulated in me when I see a tree directly before me is not to be interpreted as my knowing a correlation between two sets of ideas in my mind. The inference concerns the relation obtaining between a set of ideas in my mind and an activity which is not a set of ideas, but causes them. By the very nature of the case, I cannot at the same time perceive the tree and also perceive the links in that chain which causes me to perceive it. In fact, no layman is likely to have direct knowledge of most of the structures, nor of any of the internal processes, which are responsible for his ability to see a tree. Consequently, to give a causal account of sense perception such as that which I have given, I must be ready to accept the view that the entities and processes referred to by physiologists do exist independently of my perceiving them. And a physiologist will of course assume that those processes which occur within me when I perceive a tree are independent of any observations which *he* may make concerning them. In short, these links in the causal chain of perception must be interpreted by both the philosopher and the physiologist in a manner consonant with realism, or the whole causal analysis of perception will fail.

Similarly, to choose another example, what occurs in the medium between me and an object is not itself experienced when I experience that object. What I see is a tree, and not light rays; what I hear is a specific sound, not waves being propagated through the air. In other words, were I only to consult direct experience, what occurs in the medium would simply not exist for me at the time at which I perceive something " through " that medium. For example, if a buzzer is placed in a jar and the air is gradually exhausted, the sound which I hear will become fainter until it finally cannot be heard. If I failed to believe what physicists tell me concerning the propagation of sound waves, and if I did not take it for granted that air exists whether I perceived it or not, I would not have any way of explaining what I had witnessed: direct experience would only inform me of a diminishing sound as I watched an experimenter operate a vacuum pump. Unless, then, we take it for granted that an unperceived entity, the air, exists when it is unperceived, we would have no way of explaining what has occurred, and would not, in fact, have any good reason for believing that in this case the causal chain was interrupted through a change in the medium rather than because of a change in any other factor in that chain. A comparable realism must also be assumed in one of the strongest of all of the arguments which can be used against direct realism, the argument from the time it takes light to travel from a distant star to our eyes. The fact that a star which we see may now be extinct is a cogent reason for holding that there is a difference between our percepts and the objects which we claim to know on the basis of these percepts. Yet this argument would make no sense unless we were ready to assume that light travels at a finite velocity whether or not anyone is aware of its existence, or of its velocity, or of its capacity to stimulate the retina of a human observer. Thus, once again, we may say that the actual entities and events which are involved in the causal process of perception must be interpreted in a manner consonant with realism if we are to regard the steps which are involved in that process as evidence for the view that what is present in direct experience is not itself an independent object, but an idea which is to be distinguished from such objects.

II

In the foregoing section I have attempted to show that two of the most usual forms of argument in favor of subjectivism are not in fact adequate to establish that position. These arguments, I have contended, actually presuppose an acceptance of the existence of objects independent of perception. However, the reliance which I have placed upon our belief in such objects leads one to ask whether a belief of this kind could ever be justified. That we do have such beliefs no one has insisted more strongly than Hume himself; what he challenged was the possibility of citing any evidence to show that our belief is warranted. His challenge at this point does not consist in examining specific arguments which realists might be expected to offer; instead, he attempts to solve the matter at a stroke by showing that no adequate arguments of the sort could in fact ever be given.

In examining Hume's attempted proof of this point, we must first note that in his chapter "Of Scepticism with regard to the Senses" he poses the question which he wishes to discuss in a special manner. He asks: "What causes induce us to believe in the existence of body?"; he does not take as his point of departure the question of how we might justify such a belief, once we had adopted it. For Hume these questions were not separable, and it is easy to see that the specific answer which he gave to the first question entailed the impossibility of treating the second as independent of it. I am therefore not inclined to charge that Hume was guilty of having conflated two distinct questions; I only wish to take note of the fact that it might be possible, on some theory other than Hume's, to separate the issue of epistemological warrant from the question of the causes of our belief in the independent existence of material objects.

The second point to be noticed concerning the general structure of Hume's argument is the fact that in examining the causes of our belief in these objects he is attempting to identify the faculty or capacity of the human mind which is responsible for that belief: it is not his aim to examine the specific experiences which lead us to vouchsafe belief, or to withhold belief, regarding the independent existence of material objects. Again, I do not wish to criticize Hume for having adopted this approach; I merely wish to point out its importance for the structure of his argument. What he attempts to prove is that the senses cannot be the cause which leads us to believe in the existence of these objects, and that our reason cannot be responsible for it either. Therefore, Hume holds that its cause is to be found in what he terms the imagination. The remainder of his discussion is then given over to a rather elaborate psychological account of how the imagination, operating on what is given in experience, leads to precisely the results which he wishes to explain.

Now, it might ordinarily be thought that it is our senses which cause us to believe in the existence of body; however, Hume brings forward what he regards as decisive considerations against that view. In my opinion, these considerations are by no means decisive, for I believe them to rest on certain crucial but erroneous assumptions. These assumptions constitute the third and fourth points to which I should now like to call attention.

It will be remembered that immediately after stating that his purpose will be to explain "what causes induce us to believe in the existence of body," Hume turns his attention to a distinction between two characteristics which "body" possesses.[29] This discussion actually constitutes an implicit definition of what he is going to take the term "body" to mean. By "body" he means *objects*, and specifically those objects to which we attribute "*a continu'd existence*" and which we also suppose "to have an existence *distinct from the mind and perceptions.*"[30] By the latter phrase Hume refers both to

[29] The passage to be discussed is the second paragraph of Bk. I, Pt. IV. Sec. ii in the *Treatise*.

[30] In support of the contention that Hume has in mind specific material *objects* when he uses the term "body," I might cite the fact that when he gives his own analysis of the causes of our belief, what he uses as illustrative

external existence and to *independence of the mind.* What he therefore sets himself to demonstrate is that neither the senses nor reason
can be the causes which induce us to believe in the existence of
entities having these characteristics; rather, he holds that they are
imputed to objects by virtue of the activity of the imagination.

What I here wish to point out is the peculiar nature of Hume's
characterization of " body." With the exception of " externality," the
properties which he specifies as essential to our notion of " body "
refer only to the epistemological status of objects, and not at all to the
empirical characteristics of these objects. Furthermore, with respect
to " externality," it is to be noted that Hume cites an earlier chapter
as having proved that the question of external existence need not be
taken into account in his present discussion.[31] Therefore, what
Hume has done in this paragraph is to strip our notion of " body " of
every empirical characteristic which it contains, leaving as its residue
only the general characterization of " a something which is distinct
from the mind and which continues to exist when we are not perceiving it." The oddity of this characterization of " body " is the
third point in Hume's argument to which I wish to call attention.
And it is to be noted how important a role it plays in the structure
of that argument. Because of the way in which he identifies what
the term " body " is to be taken as meaning, Hume makes it impossible for anyone to claim that our senses could ever inform us that

---

material is our belief in such objects as doors, and stairs, and letters, and
persons, and oceans, and continents (cf. p. 196 of the *Treatise*). It is also
apparent in his use of the term " objects " in the paragraph which I am
analyzing. What he seeks to explain is why we regard these objects, and the
world which is composed of them, as " something real and durable, and as pre-
serving its existence, even when it is no longer present to my perception " (*ibid.*,
p. 197). It is to be noted how much more limited this is than the epistemological problems concerning the existence of " body " which troubled either
Descartes or Locke.

[31] The chapter cited is Bk. I, Pt. II, Sec. vi. In that chapter the Berkeley-
Hume theory of the perception of distance is not explicitly mentioned, although
it is involved. In order to avoid discussing that theory again, I shall hereafter
take " distinctness from the mind and perception " to be equivalent to " independence." This is legitimate since Hume himself remarks that " when we
talk of real distinct existences, we have commonly more in our eye their
independency than external position in place " (*Treatise*, p. 191).

body exists. In order to inform us that objects have a continuing existence, that is, that they exist when they are not being perceived, our senses would have to operate when they are not operating; in order to inform us of the existence of something which is distinct from all they contain, our senses would also have to include what, by definition, they cannot include. Thus, the implicit definition of body in terms of these general epistemological properties, rather than in terms of any directly accessible empirical properties, makes it impossible for us to say that we can find the cause of our belief in body in any data which the senses provide. To be sure, Hume is correct in holding that we do believe that bodies have a continuing existence and are independent of our perceptions of them. Where he is in error is in suggesting that this is *all* that we mean when we refer to a given entity as a material object, or body; and it may well be the case that some of these other properties are what cause us to believe that bodies have a continuing existence distinct from the mind.

We now reach the fourth point to which I wish to call attention in Hume's extremely truncated argument. In discussing the criterion of distinctness, Hume insists—and at first glance it may seem both obvious and correct—that the data given by sense are, so to speak, epistemologically neutral. As he says, "That our senses offer not their impressions as the images of something *distinct*, or *independent*, and *external*, is evident; because they convey to us nothing but a single perception, and never give us the least intimation of any thing beyond." [32] In other words, it is Hume's claim that all impressions, taken individually, stand on the same footing: no one of them more than any other can of itself cause us to believe that what we are beholding or touching or hearing is independent of us; no one of them can give us "the least intimation of anything beyond." The same point may be put in the following alternative way: according to Hume, all of our impressions appear to us to be "internal and perishing existences," none of them seeming to be more "subjective" than any others. However, Hume's contention is surely not true. So long as we are speaking of the *causes* of belief, and not of epistemological

[32] *Ibid.*, p. 189.

warrant—and this is what Hume has here set himself to do—not all of our impressions are neutral with respect to what they suggest concerning the status of that which we experience.[33]

Consider, for example, the difference between what is suggested by the appearance of a rainbow and what is suggested to us by our vision of a hill behind which a part of the rainbow's arc seems to disappear. Although we do not consider the rainbow to be merely a fleeting and perishing perception in our own mind, we nonetheless do not regard it as having the same degree of "objectivity," or permanence, or independence of all that surrounds it, as the hill appears to have. For example, while we doubtless believe that we would be able to see the rest of the rainbow's arc if the hill were not obstructing our vision, we do not suppose that the rainbow would continue to exist in the dark, as the hill presumably would. This, I submit, need not simply be the result of our sophistication with respect to physical theory: the hill and the rainbow *look* different, and do not appear as having the same form of material existence. A hill usually strikes us as being an object, "a thing," while a rainbow has the look of something intangible, as being "an appearance" in the sky. This difference between "things" and that whole class of phenomena which strike us as appearances, shadows, reflections, and

---

[33] Hume's tendency to slide back and forth between the question of the causes of a belief and the warrant for such a belief can be clearly seen in his discussion of whether the senses furnish us with the idea of something distinct from the mind. Just after saying that the senses cannot give us "the least intimation of anything beyond," Hume seeks to support his contention by the following reason: "When the mind looks farther than what immediately appears to it, its conclusions can never be put to the account of the senses." This, however, is merely to say that the senses could never provide warrant for our belief in anything which lies beyond what they present, not that they could not cause such a belief.

Precisely the same point can be seen in the sentence which closes Hume's argument against the possibility of the senses being the cause of our belief in body: "Upon the whole, then, we may conclude, that as far as the senses are judges, all perceptions are the same in the manner of their existence" (*ibid.*, p. 193). The causal question turns, however, on whether all perceptions *appear* to possess the same manner of existence, not upon whether they actually do so. If they all appeared to possess the same manner of existence, Hume would not have been able to draw his original distinction between impressions and ideas: thus it is obvious that he is here raising the question of epistemological warrant.

the like, is one of the most familiar distinctions which we draw within our experience, regardless of how it arose. Hume would of course hold that such a distinction depends upon custom, or, more precisely, that it arises through the action of the imagination operating on whatever constancies and coherences were given in past experience, or were suggested by it. On the other hand, there seems to be ample evidence to show that independently of our previous experience, some of the qualities which we perceive suggest permanence and stability, whereas others do not.[34] For example, as we have previously noted, there appears to be this kind of difference between the "reality-character" of surface-colors and of film-colors. Furthermore, anything which appears shadowy, or which seems to lack depth and bulk, seems "unreal," or seems "less real" than do other objects. And objects with characteristics such as sharply defined contours, three-dimensionality, and a discernible micro-structure in their surfaces, usually appear stable and "real." Furthermore, it is important to note that those material objects which appear "real" also appear to be independent of us, possessing an apparently continuing existence which marks them off as distinct from the fluctuations of our attention. Therefore, had Hume not stripped the notion of "body" of all of those specific empirical qualities which what we call material objects seem to possess, he might have found that some of these qualities, in contrast to others, do actually suggest that the objects which possess them have a continuing existence, and are therefore regarded as distinct from our minds.[35] For example, when

[34] For a theoretical discussion of this problem and for supporting experimental findings, cf. A. Michotte, "À propos de la permanence phénoménale," and the monographs by Sampaïo and Knops which are there cited. Also, cf. A. Michotte, "Le caractère de 'réalité' des projections cinématographiques."

Another way of suggesting the inadequacy of Hume's type of explanation of the "object-character" of some of our perceptions is to ask how he could give a genetic account adequate to explain the figure-ground differentiation which exists in "nonsense-figures" no less than in the perception of familiar objects. Similarly, we might ask what genetic account he would be able to give to explain the principles of camouflage.

[35] In Hume's psychological account of the matter, it is our belief in continuing existence which underlies our belief in distinctness, and not the reverse (Treatise, p. 199). In so far as distinctness is to be identified with

we have impressions of entities which appear to be steady rather than fluctuating, which appear stable rather than evanescent, which are vivid and precise in detail, rather than filmy and vague, we are likely to attribute a permanence to them, and this permanence is not to be found in our own inner states of consciousness. It is such impressions which suggest to us that what we see or hear or touch is not merely an internal and perishing existence, but an independent object with which we have come into physical contact.[36]

While Hume never calls attention to these striking differences among our perceptions, his own explanation of why we believe in the existence of body covertly makes use of them. In fact, if such differences were not taken for granted by a reader, Hume's account of our belief in the independent existence of objects would not possess the plausibility that it does. Consider, for example, the following passage which constitutes a summary preview of his position:

> When we have been accustom'd to observe a constancy in certain impressions, and have found, that the perception of the sun or ocean, for instance, returns upon us after an absence or annihilation with like parts and in a like order, as at its first appearance, we are not apt to regard these perceptions as different (which they

"independence," this seems to me true. However, in so far as the notion of distinctness also includes external localization, I should of course hold that it is false.

[36] An equally sharp contrast between what appears as "without" and what appears as "within" is not to be found when we perceive fleeting external events; consequently, in the latter cases we are likely to feel less sure that what we saw or heard was really an external and independent event, rather than a mere idea in our own minds. However, even among fleeting events there are differences in the "objectivity" which they appear to possess, and this difference is attributable to factors such as whether they possess sharply demarcated "contours." (Compare for example the difference between a streak of forked lightning and a flash of heat-lightning.) To put the matter crudely, some events are more "object-like" than others, even when they are of an extremely short duration. It appears to me that the more an event possesses this characteristic, the less likely we are to doubt its independent existence. Furthermore, it seems to me that this characteristic may be part of what Hume wished to include when he referred to "vivacity," although he himself seems to have regarded that attribute of our impressions as being not further analyzable, and as not being dependent upon the specific nature of the impression but only upon its "impact."

really are) but on the contrary consider them as individually the same, upon account of their resemblance.[37]

In this passage, I submit, the whole plausibility of Hume's account rests on his use of objects such as the sun and the ocean as illustrations. The "return upon us with like parts and in a like order" of a story which we have heard, of a dream which frequently recurs, of a mathematical proof which we remember, would not in the least suggest to us that the story, dream, or proof possesses a continuing and independent existence. What Hume is taking for granted is that we all know what objects are, and that we can all distinguish between ideas which are only in our minds and impressions which come to us from something external to us. All that the constancy and coherence of our experience is used to explain is how we come to attribute a continuing and independent existence to that which we have already experienced as "real," as distinct from something merely "subjective." To be sure, it might be thought that Hume could then go on to explain in precisely the same terms why we believe that a particular congeries of perceptions constitutes a single object; however to explain in terms of constancy and coherence what makes the sun or the ocean appear to us as an object is a task which Hume did not actually attempt, and it is one which would not have been easy of fulfilment.[38] In short, Hume's explanation of our belief in the continuing and independent existence of objects seems to me to borrow whatever plausibility it contains from the fact that the objects which Hume chose to cite were entitities which evoked that belief for reasons other than those which he actually employed in his explanation. And these other reasons, I have suggested, involve the way in which our perceptions appear to us in immediate experience, rather than being a result of the way in which Hume holds that the imagination functions.

The upshot of the preceding argument is obvious: the cause of

[37] *Treatise*, p. 199.

[38] Berkeley, and perhaps also Locke, can be criticized on the same grounds. However, the criticism seems to me to be more damaging when directed against Hume, since he purports to be offering a psychological explanation which, as part of his science of human nature, is to be regarded as an empirically verifiable genetic account.

our belief in the continuing existence of an independent world of objects is to be found in our immediate perceptual experience, that is, in the senses.[39] This is not of course to say that such a belief cannot be challenged. However, what Hume originally set out to explain was why in point of fact we do believe that there are objects which have a continuing existence and which exist independently of us; the question of what epistemological warrant such a belief may have is a separate and independent question, once we have given an answer other than Hume's to the causal question. In seeking to assess the warrant of that belief one would have to cite reasons for either accepting or rejecting it. In Hume's opinion an examination of the relevant arguments inevitably leads to skepticism, and he held that the more we trust to reason in this matter the more virulent our skepticism will become. However, as we saw in the initial section of this chapter, he erroneously thought that the two classic arguments against a naïve or direct realism could serve to undermine realism itself. Our consideration of these arguments should have served to suggest that inquiry supports our natural realism even while it forces us to alter some of our beliefs as to what characteristics are possessed by that which we perceive.

III

At this point there still remains one feature of Hume's philosophic position which might be thought to prove that it is impossible to escape some form of skepticism regarding the independent existence of the physical world. This feature is his attempt to remain within the confines of what he regarded as a pure empiricism, according to which all human knowledge depends upon experience, and experi-

[39] It should perhaps be pointed out that what is given in immediate perceptual experience is not to be identified with what is reflected on the peripheral sense organs: the retina, for example, is but one link in the causal chain in perception, and what occurs on the retina should not be considered to be identical with "the given" any more than one identifies the given with those processes which occur in the optic nerve. To be sure, the doctrine that "perception equals sensations plus implicit inference" has frequently led to confusion on this point, since "sensation" has often been identified with whatever occurs on the periphery of the body.

ence consists in what is directly present to the mind. On the basis of such a postulate it would seem impossible to escape subjectivism. To be sure, if one were to present a full exposition of Hume's assumptions and his aims, my statement of his empirical starting point would demand not only careful elucidation, but some qualification; however, such an exposition is not my present concern. I only wish to show that Hume's initial and basic postulate concerning the materials of human knowledge (however one phrases that postulate) itself presupposes a tacit acceptance of epistemological realism, and that this should have deterred Hume from using this postulate as a means of buttressing his other arguments for subjectivism. Put more concretely, I shall show that in Hume's original distinction between impressions and ideas—a distinction integral to the whole of his theory of knowledge—a necessary realistic assumption is already contained.

It is in the opening of the first chapter of the *Treatise* that Hume draws his distinction between our impressions and our ideas. He says:

> All the perceptions of the human mind resolve themselves into two distinct kinds, which I shall call *impressions and ideas*. The difference betwixt these consists in the degree of force and liveliness, with which they strike upon the mind, and make their way into our thought or consciousness. Those perceptions which enter with most force and violence, we may name *impressions*; and, under this name, I comprehend all our sensations, passions, and emotions, as they make their first appearance in the soul. By *ideas*, I mean the faint images of these in thinking and reasoning; such as, for instance, are all the perceptions excited by the present discourse, excepting only those which arise from the sight and touch, and excepting the immediate pleasure or uneasiness it may occasion.

Now, as has frequently been pointed out, "force and liveliness" are not in fact sufficient to enable us to distinguish between some of our ideas and some of our impressions: for example, the memory of a *faux pas* which we have committed may be a far more lively perception to us than are the sights and sounds which surround us.

And cases of this sort are immediately admitted by Hume.[40] He insists nonetheless that we can all adequately distinguish between "feeling and thinking." He tends to suggest that our ability to make this distinction depends in part on the order in which impressions and ideas are experienced, our ideas coming after certain impressions from which they are copied, or out of which they have been composed. However, this second criterion for the distinction between feeling and thinking will also not do. In the first place, there is the well-known difficulty as to how Hume, on the basis of his theory of memory, can account for comparisons between two perceptions such that one of them can be recognized as being the memory-image of the other. In the second place, the "copying" criterion for distinguishing between thinking and feeling would result in the oddity that every perception which did not strike us as resembling some earlier perception would be identified by us as an impression, and not as an idea; and, further, that every perception which did in fact strike us as resembling some earlier perception might be considered to be an idea, rather than in some cases being considered a new impression. Each of these consequences is, of course, out of line with our experience, for we sometimes have perceptions which we regard as ideas even though we find ourselves unable to recall the impressions from which they were drawn, and it is also the case that we may see a familiar sight and recognize it as a sight that we had seen before, without being in the least confused as to whether we are seeing it or are merely remembering or imagining it. In the third place, it is obvious that Hume himself drew the distinction between impressions and ideas on some basis other than their temporal order, for in the famous instance of the missing shade of blue he acknowledged that we might have a simple idea without first having had a simple impression from which it was copied. And this is even clearer with respect to our complex ideas of imagination, which Hume allowed that we can form more or less at will. Though these complex

---

[40] That he distinguishes between impressions and ideas even when a difference in their vivacity is not experienced is not only clear in this passage but is also to be found in his discussion of impressions and ideas in Bk. II, Pt. I, Sec. xi (pp. 318–19 of the *Treatise*).

ideas are not necessarily copied from complex impressions, we do know that they are in fact ideas, and not impressions. For example, in the case given by Hume, we surely know that when we are envisioning "the New Jerusalem, whose pavement is gold, and walls are rubies," that what we are experiencing is an instance of imagining, and is not a case of having an impression. All three of these sorts of reasons show that the distinction between impressions and ideas, or between feeling and thinking, cannot be based on the temporal sequence in which our perceptions occur. And, as we have seen, Hume also admits that the distinction cannot be based on vivacity alone. On what, then, is it based? I submit that it is actually based on the role which our sense organs play in our impressions, a role which they do not play when our perceptions are to be classified as ideas.[41]

To be sure, Hume specifically disavows such an interpretation, for he claims not to be treating any of our perceptions in terms of their causes, and he seeks to regard them as what might best be designated as "contents of consciousness."[42] Nonetheless, a careful reading of Hume demonstrates that he in fact constantly thinks of impressions as arising from physical causes operating on our senses. For example, at the very outset of the *Treatise*, where he admits that although "in sleep, in fever, in madness, or in any very violent emotions of soul," our ideas sometimes approach our impressions with respect to vivacity, he nonetheless refuses to acknowledge that when we are in these states our perceptions actually are impressions; rather, he

---

[41] I am here only attempting to say what *Hume's* distinction is actually based upon; a positive consideration of the question would take us unnecessarily far afield.

[42] Cf. his terminological footnote where he says:

"I here make use of these terms, *impression* and *idea*, in a sense different from what is usual, and I hope this liberty will be allowed me. Perhaps I rather restore the word idea to its original sense, from which Mr. Locke had perverted it, in making it stand for all our perceptions. By the term of impression, I would not be understood to express the manner in which our lively perceptions are produced in the soul, but merely the perceptions themselves; for which there is no particular name, either in the English or any other language that I know of " (*Treatise*, Bk. I, Pt. I, Sec. 1).

regards them as ideas.[43] However, if we are to determine what leads him to do so, we must inquire what these four types of state have in common. If we were to conduct such an inquiry from the point of view of what is immediately given within consciousness when we ourselves dream, or have a fever, or experience violent emotions (let alone what would occur were we mad), I very much doubt whether we could find any common denominator. Nevertheless, it is surely not accidental that Hume should have classed these states together. In doing so he was, I suggest, observing them as an outsider, and not in terms of whatever perceptions occur within them. Examining them in this way, it seems clear that sleep and fever have in common the fact that in them our bodies are so affected that what we are aware of is regarded by others (or by ourselves at a later time) as having been internally engendered. And whether or not Hume believed that madness might also be related to specific bodily states, what is taken as a symptom of mental derangement is the fact that the madman hallucinates—that what are apparently taken by him to be a set of impressions given through the senses are in fact ideas which have no such causes. And the very violent emotions of soul which Hume might here have had in mind may surely also be cases in which an outsider takes note of the fact that what an enraged person apparently experiences bears little relation to what, were he in another state, he presumably would experience. Similarly, we may note that in this same first section of the *Treatise*, Hume distinguishes between " the *idea* of red, which we form in the dark, and that *impression* which strikes our eyes in sunshine " (my italics). And in the immediately following section he abandons his original terminology in which impressions were spoken of as " striking upon the mind," using instead the locution that impressions " strike upon the senses." Precisely the same point is also clear in the opening paragraph of the *Enquiry*, where memory and imagination are spoken of as mimicking or copying " the perceptions of the senses " (i. e., our impressions), and where Hume rephrases the dictum of the *Treatise* concerning the difference between feeling and thinking by saying " the most

---

[43] In Section II of the *Enquiry* he also makes an exception of those cases in which " the mind [is] disordered by disease or madness."

lively thought is still inferior to the dullest sensation." Time after time any attentive reader will see that Hume is in fact appealing to a causal theory of perception to explicate the difference between impressions and ideas.[44]

To this contention it might be objected that Hume does not confine his use of the term "impressions" to such perceptions as are assumed by him to be caused by the action of our sense organs. Yet an examination of the two passages in which he draws a distinction between those impressions which depend upon our senses and those which do not, clearly shows how fundamental a role Hume assigned to the action of the senses in all of our experience.[45] In these passages he distinguishes between "impressions of sensation" and "impressions of reflection," the latter term being applied to those states of mind which he designates as passions. However, it is significant that unlike Locke and Berkeley, Hume insisted that all of these states of mind were complex derivatives of impressions of sensation: for him there could be no experience which did not include sensory elements. To be sure, not all of these sensory elements came through the peripheral sense organs: within the class of our "impressions of sensation" Hume also included sensations of pleasure and pain. Unfortunately, he was less clear than one might wish concerning what relations obtain between these two types of sensation. However, what is of interest for us to note is that even those sensations which do not come to us through our peripheral sense organs were considered by Hume to be dependent upon causes which lie outside of our immediate experience. As he says:

---

[44] For further quotations bearing on this point, cf. Sterling P. Lamprecht "Empiricism and Epistemology in David Hume," in *Studies in the History of Ideas*, II, 222–25, *passim*. While Lamprecht does not make precisely the point which I am making, his discussion is relevant to my point, and he does say that "like most writers of the century, Hume took up the problem of the source whence come our first impressions of sensation." While admitting that Hume is not consistent in doing so, he stresses the extent to which "Hume regards perceptions as mental existences caused by a nonmental external something" (p. 225). The same inconsistency is pointed out by T. H. Green in Section 201 of his Introduction to Hume's *Treatise* (cf. Green, *Works*, I, 166–68).

[45] The two passages under consideration are *Treatise*, Bk. I, Pt. I, Sec. ii, and Bk. II, Pt. I, Sec. i.

Original impressions or impressions of sensation are such as without any antecedent perception arise in the soul, from the constitution of the body, from the animal spirits, or from the application of objects to the external organs. Secondary, or reflective impressions are such as proceed from some of these original ones, either immediately or by the interposition of its idea. Of the first kind are all the impressions of the senses, and all bodily pains and pleasures: Of the second are the passions, and other emotions resembling them.[46]

In other words, what characterizes all original impressions—including those which are "inner sensations"—is the fact that they must be attributed to causes which lie outside of experience. "Impressions of reflection," on the other hand, are considered secondary by Hume

[46] *Ibid.*, Sec. i.

It is to be noted that in the earlier parallel passage (Bk. I, Pt. I, Sec. ii) Hume's classification of our impressions of sensation included not only "the impressions of the senses" (plus "all bodily pains and pleasures"), but also included such other "inner-sensations" (or "body-sensations") as thirst-or-hunger and warmth-or-cold. (By the latter Hume clearly meant a sensation related to our own bodily feelings, such as "*being* warm or cold," rather than our experience of feeling *that* a radiator is warm, or a lump of ice is cold.) Hume apparently believed that such inner-sensations are like impressions of the senses except that they were caused by the actions of our own internal organs, without the necessity of a stimulus affecting our peripheral sense organs. For example, he says: "Hunger arises internally, without the concurrence of any external object." And in this same passage he seems to imply that it, as well as some other impressions, can be produced from "within" and are attributable to "organs [which exert themselves] like the heart and arteries, by an original internal movement" (*ibid.*, Bk. II, Pt. I, Sec. v, [p. 287]). This is both clear and intelligible, but it is difficult to see how Hume could distinguish between such body-sensations and our other impressions except in terms of a contrast between what appears as outside of us, and what is localized as occurring within our own bodies. However, on Hume's theory of how we come to form the idea of external existence, such a distinction has to be derived from past experience, whereas Hume here supposes that we know from the outset the difference between "outward" and "inward" sensations. (He uses these terms in the last paragraph of Section II of the *Enquiry*.) What remains most unclear throughout Hume's discussion of our inward sensations is how the sensations of pleasure and pain are related to our external sensations, and to our other internal sensations. While this is not an important defect in his theory of knowledge, I believe that it raises a whole series of difficulties for his analysis of the passions, and for his analysis of our moral notions.

precisely because they can be traced back to our original impressions. Thus, as I have insisted, Hume actually used causal criteria throughout his classification of the elements within our experience. This point may be further illustrated through noting why Hume employed the term "impressions" to refer to our passions. He did so in order to point out that there is also a difference between actually experiencing an emotion and merely recalling it. Although he wished to distinguish between these experiences by using the criteria of vivacity and of precedence,[47] the real criterion by means of which he identified what constitutes an impression of reflection was whether that perception included as one of its elements sensations of pleasure or of pain. As he says: "Thus pride is a pleasant sensation, and humility a painful; and upon removal of the pleasure and pain, there is in reality no pride and humility. Of this our feeling convinces us; and beyond our feeling, 'tis here in vain to reason and dispute."[48] This, however, signifies that even our impressions of reflection are only to be designated as impressions because of their sensory components, that is because they belong to that class of experiences which "arise in the soul from the constitution of the body, from the animal spirits, or from the application of objects to the external organs." That Hume has no right to speak in this way is of course true, but that is not the point which I here wish to make. What I have been concerned to show is that in all cases in which Hume speaks of "impressions" as distinct from "ideas," his distinction between these two great classes of perceptions rests on the fact that the former have as their causes something which is independent of any relations which we can trace within our own consciousness. And, finally, it is to be noted that since neither our sensations of pleasure and pain nor our passions can be taken as providing us with data concerning matters of fact, those impressions upon which all

[47] In the *Enquiry* the criterion of vivacity was stressed by Hume (cf. Section II, pp. 17–18); on the other hand, in both of the passages which I have cited from the *Treatise*, the main emphasis is laid on the precedence of our impressions.

[48] *Treatise*, Bk. II, Pt. I, Sec. v (p. 286). Cf. also Bk. I, Pt. I, Sec. ii where the return of pleasure and pain upon the soul is held to lie at the basis of all impressions of reflection.

human knowledge is founded are identified by Hume with the data given us through our external sense organs.[49]

Now, it must again be pointed out that Hume specifically disavows any interest in tracing the causes of our original impressions, whether they come to us through the senses or whether they are inner sensations. Nonetheless, as we have seen, he clearly does believe that all such impressions result from the operation of hidden causes. Furthermore, in both of the passages which concern the distinction between impressions of sensation and of reflection he suggests that an examination of these causes belongs to the province of anatomy and natural philosophy. Presumably, then, there are empirical means by which we can establish some of the causal factors involved in impressions, even though Hume himself does not envision it as his task to do so. However, in opposition to this thesis we must now note

[49] It is worth noting that Berkeley used precisely the same criterion of a sensory origin to distinguish among our ideas, and that he too regarded all of our knowledge (except that which concerned minds) as being attributable to our senses. We may note the presence of these views at the very foundation of his system. For example, item #378 in his *Philosophical Commentaries* reads as follows:

1 All significant words stand for Ideas
2 All knowledge about our ideas
3 All ideas come from without or from within.
4 If from without it must be by the senses & they are called sensations.

This item is identified by A. A. Luce, the most recent editor of the Notebooks (and from whose edition I quote), as "the demonstration of the New Principle . . . and the doctrinal climax of Notebook B" (*The Works of George Berkeley, Bishop of Cloyne*, I, 123). In the same connection we may cite the opening of item #539 in Notebook A: "ffoolish in men to despise the senses. if it were not . . . yᵉ mind could have no knowledge no thought at all." And Hume's distinction between "impressions" and "ideas" parallels Berkeley's first and third classes of the elements of knowledge, as given in *The Principles of Human Knowledge*. Berkeley opens Part I of that work by saying:

It is evident to any one who takes a survey of the objects of human knowledge, that they are either *ideas actually imprinted on the senses*, or else such as are perceived by attending to the passions and operations of the mind, or lastly *ideas formed by help of memory and imagination* . . . . [my italics].

It is obvious, then, that Berkeley too assumed that we can distinguish without difficulty between what is given to us through a sensory experience and what depends upon the activity of memory and imagination.

that there are other passages in which Hume insists that it is in principle impossible to know anything concerning the nature of what gives rise to our perceptions. The most famous of these statements is probably the following passage from the *Enquiry*:

> By what argument can it be proved, that the perceptions of the mind must be caused by external objects, entirely different from them, though resembling them (if that be possible) and could not arise either from the energy of the mind itself, or from the suggestion of some invisible and unknown spirit, or from some other cause still more unknown to us?

And to this Hume answers:

> It is a question of fact, whether the perceptions of the senses be produced by external objects, resembling them: how shall this question be determined? By experience surely; as all other questions of a like nature. But here experience is, and must be entirely silent. The mind has never anything present to it but the perceptions, and cannot possibly reach any experience of their connexion with objects. The supposition of such a connexion is, therefore, without any foundation in reasoning.[50]

Yet were this conclusion true, Hume would have had no right to draw the distinction which he originally drew between our impressions and our ideas, since that distinction tacitly assumed that whatever might be the nature of the causes of our impressions, those impressions did in fact have causes lying outside the sphere of our immediate experience, while all of our ideas were traceable to the effects of the mind's operation on these prior impressions. And if this distinction between impressions and ideas were to collapse, Hume's empirical theory of knowledge would also collapse, for it rested on that distinction: in essence, it consisted in the requirement that we should ascertain the meaning of a term by tracing our idea of it back to the impressions from which it was derived.[51]

I do not wish to be understood as contending that Hume could not consistently hold to his subjectivistic position: once he had reached

---

[50] *Enquiry*, Sec. XII, Pt. I, pp. 152–53.
[51] Cf. the conclusion of Sec. II of the *Enquiry* (p. 22).

that position, there is no doubt whatsoever that he could retranslate his distinction between impressions and ideas into subjectivistic terms. He could, for example, allow anatomists and natural philosophers to explain the causes of our sensations, and he could then translate all that they have to say concerning light waves and the retina and our animal spirits into subjectivistic terms. However, as we saw in the first section of this essay, such an account would lose its evidential foundation if we took seriously the translation which a subjectivist would have to make of these terms. And as we have now seen, Hume's own empirical starting point, his very insistence that knowledge depends upon sensory experience, would also lose its plausibility once his subjectivistic thesis were allowed to undercut our untutored belief that sensory experience involves our being affected by independently existing material objects which act upon our sense organs. Therefore it is my contention that Hume's system collapses not due to any logical contradiction which it unavoidably contains, but because the conclusion which it presumably reaches does not rest on arguments which are in the least plausible if in the end it turns out that we must take that conclusion to be true.

As one further illustration of this point, consider Hume's contention that the senses cannot provide us with any reasons to believe that objects exist independently of us. In that argument he insisted that all perceptions of whatever kind must stand on the same footing, a point which he had also made in his earlier chapter on our idea of external existence, where he said: "To hate, to love, to think, to feel, to see; all this is nothing but to perceive." [52] But this is precisely what he denies in his distinction between impressions of sensation and impressions of reflection: in drawing that distinction he assumed that to hate and to love have an entirely different set of causes than seeing. Were subjectivism to be taken as true, we should have no reason to suppose that there was any such systematic distinction among our various impressions, and Hume would not therefore have the right to assume—as in truth he does assume—that we immediately know the difference between impressions and ideas, and even between those impressions which only convey data concerning

[52] *Treatise*, Bk. I, Pt. II, Sec. vi (p. 67).

our feelings and those which are caused "by the application of objects to the external senses."

As a final illustration of the same point, I should now like to show that if we take the subjectivistic thesis to be true, the thesis which Hume recognized to be central to his whole theory of knowledge loses the support of those arguments by means of which he sought to establish it. That central thesis Hume summarized in the proposition that "All of our simple ideas in their first appearance are derived from simple impressions."[53] Hume attempted to demonstrate the truth of this contention by two arguments: (1) "To give a child an idea of scarlet or orange, of sweet or bitter, I present the objects, or in other words convey to him these impressions; but proceed not so absurdly, as to endeavour to produce the impressions by exciting the ideas"; and (2) "Where-ever by any accident the faculties, which give rise to any impressions, are obstructed in their operations, as when one is born blind or deaf; not only the impressions are lost, but also their correspondent ideas." Now suppose that we attempt to interpret these two extremely plausible arguments in terms of the subjectivist thesis: they immediately lose what plausibility they had. In the first place, we must not only translate "the child" and "those objects which we present to him" into sets of perceptions in our own minds, but we must also find some way of gaining access to what we would ordinarily call *his* inner experience in order to be able to observe the sequences in which it unfolds. How we could achieve the latter feat is hard to imagine, if it is true that all we can know are our own perceptions. A similar difficulty also obtains when we attempt to establish the second argument concerning those who are born deaf or blind. However, since some may think that it is no more difficult to explain this on the basis of subjectivism than it is to do so on the basis of realistic assumptions, I shall forego discussing this aspect of Hume's arguments further.

There is, however, another aspect of both of these illustrations which is worth noting, although Hume does not call attention to it. In both of them there is implicit the assumption that there not only is a necessary order with respect to the relations between our impres-

[53] *Ibid.*, Pt. I, Sec. i (p. 4).

sions and our ideas, but that there also are, apparently, some elements of necessity with respect to the relations which impressions bear to each other.[54] This can be seen in a further illustration which Hume uses in support of his two arguments, and which I prefer to discuss since it raises no questions concerning other minds.[55] Hume says: "We cannot form to ourselves a just idea of the taste of a pine-apple, without having actually tasted it." This statement is intended to provide one additional illustration of the precedence of our simple impressions over any of our ideas. Yet, actually, before it can be used to prove this contention it must be interpreted as showing that in our experience there actually are some elements of necessity regarding the order in which our impressions are given. For what Hume's illustration presupposes is that any new impression of taste will only arise in our consciousness when accompanied by some other impressions belonging to our senses of sight and of touch. To be "actually tasting" the pineapple means, on the subjectivist thesis, to be having a set of impressions of our own bodily organs and of other objects, such that all of these perceptions form that coherent pattern which is known as being awake and sensing something, rather than imagining it or dreaming it. Thus, our new impression of this particular taste is assumed by Hume to be necessarily connected with other impressions.[56] Such an orderly connection among our impressions can of course be reinterpreted in terms of the subjectivist posi-

[54] By "necessity" I do not wish to be taken as referring to logical necessity: it is not the case that experience even suggests that there is a connection of logical necessity among our impressions. Although Hume and others often do equate "necessity" with "logical necessity," the term can also be used to refer to invariant connections over which we do not have control. It is in that sense, and that sense only, that I am here using it.

[55] The point which I am about to develop can also be made with respect to each of the two arguments which Hume used, once one has solved the question—or begged the question—of how, on the subjectivist thesis, we may be interpreted as knowing what transpires in other minds.

[56] Of course, the taste need not be connected with any one other single impression, for we can taste something without first having seen it, and we can taste it without having touched it with our fingers, etc. However, Hume would insist that we cannot experience its taste without also being in a position to experience at least some of its other qualities, for this is precisely what is involved in speaking of the taste as being an impression rather than merely being an idea.

tion; however, in terms of that position one could not offer any explanation of why such elements of order constantly appear and reappear within our experience. Yet Hume's distinction between impressions and ideas ultimately rests on assuming that the order which obtains among those perceptions which are to be denominated as impressions is not an order which we can control, and it is an order which is attributed by him to "unknown causes" which act on our senses. The less weight he would wish to attach to there being such causes for the elements of inescapable order which we find among our perceptions, the more tenuous would become his distinction between impressions and ideas. Conversely, the more weight Hume attached to the empiricist thesis that all knowledge has its source in impressions, the more need there was for him to hold that the order which we find among some of our perceptions is an order which has a cause outside of ourselves.[57]

In short, it seems to me that a consistent subjectivism undermines the plausibility of the thesis that we all know the difference between actually seeing something and merely thinking about it: the more one stresses the thesis that all we know are our own states of mind, the harder pressed will one be to make clear wherein the difference between seeing and thinking really lies. And the more sure we are that we can tell this, the less plausible subjectivism will seem to us to be.

The immediately preceding conclusion is of course related to the conclusion which was reached in Section I of this chapter when the two classic arguments against epistemological realism were examined.

[57] It is to be noted that Berkeley, who also started from the distinction between what comes to us through the senses and what does not (cf. note 49, above), distinguishes between these contents of consciousness by means of elements of order for which we are not ourselves responsible. For example, in *The Principles of Human Knowledge* (#30) he says: "The ideas of sense are more strong, lively, and distinct than those of imagination; they have likewise a steadiness, order, and coherence, and are not excited at random, as those which are the effects of human wills often are, but in a regular train or series." Similarly in #33 he links regularity and constancy with vividness and distinctness, and links orderliness and coherence with strength, as criteria of what are to be termed "real things."

Both of those arguments, as I attempted to show, did in fact presuppose the acceptance of some form of realism if they were to be regarded as well-founded. Thus, they could not legitimately be used as a means of refuting realism. We have now seen that Hume's distinction between impressions and ideas also involved realistic assumptions concerning the causes of our impressions; as a consequence, whatever aspects of Hume's system presuppose that distinction are aspects which cannot legitimately be used as a means of discrediting epistemological realism. Since there is little in Hume that concerns realism and subjectivism which does not involve a reliance upon his distinction between impressions and ideas, I feel doubly secure in my contention that his grounds for rejecting realism were inadequate. Yet this conclusion could scarcely carry conviction if one found oneself forced to accept Hume's analysis of the causes operative in our belief in an independent external world. For that reason it was also necessary to show—in Section II—wherein that analysis may be said to have failed.

4

TOWARD A
CRITICAL REALISM

IN THE PRECEDING CHAPTER, using Hume as an example, it was my
purpose to show that two of the standard forms of argument for the
subjectivist thesis will not bear the weight which is frequently
placed upon them. Furthermore, I examined other aspects of Hume's
remarkable and subtle system in order to show that at these points,
too, he had failed to offer reasons sufficient to discredit epistemological realism. In these arguments there were several places at which
a reader might have noted similarities between my objections to the
subjectivistic account of our experience and some of the objections
which Gilbert Ryle has used in the same connection.[1] Now, however, we have come to a point at which the issue is not one of
whether subjectivism is capable of discrediting realism, but of inquiring as to what form an adequate realism must possess. In my
appraisal of the two classic subjectivistic arguments I did not suggest that these arguments were ineffectual if they were taken as
disproofs of some forms of realism, instead of being regarded as
means of establishing subjectivism; actually, I believe that they offer
important reasons for rejecting the characteristic theses of direct or
naïve realism. With this opinion Ryle would of course disagree.
Furthermore, I differ from Ryle in holding that many of the results
of a scientific inquiry have an important bearing on epistemological
issues. While this conviction may already have become apparent
through my discussions of Locke, Newton, and Boyle, in the present
chapter I shall attempt to establish it as firmly as I can. What I
wish to show is that the fruits of scientific inquiries, far from being

[1] Cf. *Dilemmas*, Ch. 7, especially pp. 94–96.

171

epistemologically neutral, should lead us to adopt a radical form of critical realism.

As a preliminary step toward this more positive discussion, I should like to examine with considerable care the form of direct realism which is maintained by Professor Ryle.[2]

I

As we have noted with respect to the phenomenology of perceptual experience, there are many cases in which what we perceive strikes us as being wholly independent of us, and independent also of the particular conditions under which perception takes place. This is likely to be true, for example, when we look at objects such as books, or chairs, or trees, and see them under what we regard as normal conditions of vision. It is in cases of this sort that direct realism appears to be the only epistemological position which is in accord with ordinary language, and with common-sense interpretations of our experience. And it is to cases of this sort that G. E. Moore and Ryle, among others, have made constant appeal.

Now, there are various ways in which one might attack a realism that seeks to rely upon such cases. One might, as I have noted, rephrase most of the classic arguments from contradictions and from the causal chain involved in perceiving, and show that while these arguments do not establish subjectivism they *do* serve to disprove direct realism. Or, one might offer straightforward scientific accounts of the actual, detailed processes involved in the various forms of sense perception, and examine whether the claims of direct realists are in fact compatible with these accounts. This, I think, would ultimately be the more successful method, and we have one excellent recent example of its use in Dr. Colin Strang's article, "The Perception of Heat," in the *Proceedings of the Aristotelian Society* for 1961. However, there are two special difficulties which such a

---

[2] Some of what immediately follows is drawn from an address to the Eastern Division of the American Philosophical Association, given in December, 1962, and printed in volume XXXVI (1962–63) of the *Proceedings* of that Association.

method faces when one's aim is to criticize not the theory of direct realism as such but specific formulations of it, such as one finds in Ryle or in Moore.

In the first place these realists have couched their theories in extremely general terms, and do not say precisely what they are willing to affirm, and what they would wish to deny, with respect to the nature of material objects. For example, while Ryle does occasionally mention the distinction which modern philosophers have drawn between primary and secondary qualities, what he says fails to tell us whether there actually is a distinction to be drawn between them, and, if there is, in what it consists. He says: "Secondary Qualities are not subjective, though it remains true that in the country of the blind adjectives of colour would have no use, while adjectives of shape, size, distance, direction of motion and so on would have the uses that they have in England." [3] This statement, even in its full context, tells us little, for it only involves a contrast between the experience of those who are blind and those who are not, and this is not an issue which has been involved in any discussions of primary and secondary qualities. What Ryle fails to state is whether there is any difference between adjectives referring to colors and adjectives referring to shape, size, distance, and direction of motion, with respect to how these adjectives are to be used in characterizing objects which *cannot* be directly perceived. For example, in referring to two things which existed on the surface of the earth before any sentient being existed, some philosophers would not regard it as correct to say that these objects *were* brown and grey, respectively; instead, they would hold that we should say of these objects that they would have looked brown and grey to us had we been there (or that we would have described them as being brown and grey). On the other hand, they would be willing to say that the two objects were at a certain distance from each other, and they

---

[3] *The Concept of Mind*, p. 221. It is unfortunate that in this passage Ryle did not consider various alternative uses of the term "subjective"; for, if he had, he would have noted that few philosophers have held that the secondary qualities are "subjective" in the sense of that term which he discusses.

For another brief and also unclear reference to the same problem, cf. Ryle, *Dilemmas*, p. 84 f.

would not regard it as necessary to use the locution that, had we been there, we would have described them as being that distance apart. Now I do not wish to ask whether, when one refers to an object which no one could perceive, there are good reasons for distinguishing between those qualities which such an object *did have* and those qualities which it *would have had* had we been able to perceive it; I only wish to point out that this is a question which Ryle has not discussed. I must add, furthermore, that one cannot regard it as a question which should be immediately dismissed as either unmeaningful or inconsequential; the issue at stake is not simply a question of what descriptions we are to give of material objects, but of how we are to explain the action of material objects upon one another, and, more especially, how we are to explain their action upon us when we perceive them. It is this issue which underlies the classic distinction between primary and secondary qualities.[4] In their avoidance of the problem Ryle and also Moore manage to avoid the causal problem; or perhaps one should say that in avoiding all discussions of the causal problem they avoid the problem of whether all perceived qualities of objects exist in these objects independently of perception. In either case, Strang's admirably specific and detailed approach is not one which can readily be used in showing in what respects their arguments fail.

There is a second reason why Strang's method of arguing would not be helpful in convincing those who adopt the forms of direct realism which are represented by either Moore or Ryle. This reason is related to the views which each of the latter holds with respect to the scope of psychology. According to his own account of the matter, Moore's psychological convictions were partly formed under the influence of Ward, Stout, and James, and his approach to the question of perception shows the mentalistic bias which was characteristic of Ward and Stout, in contrast to James.[5] Such a bias is,

---

[4] Cf. above, pp. 18 f.

[5] For Moore's statement, cf. *The Philosophy of G. E. Moore*, p. 29.

In Ward's *Encyclopedia Britannica* article, "Psychology," which was written in 1885 for the Ninth Edition and was subsequently twice republished, Ward wrote: "Of all the facts with which he deals, the psychologist may truly say that their *esse* is *percipi*, inasmuch as all his facts are facts of presentation,

for example, quite clear in the following passage from the lectures which Moore gave in 1910–11, and which are now published as *Some Main Problems of Philosophy*:

> The occurrence which I mean here to analyze is merely the *mental* occurrence—the act of consciousness—which we call *seeing*. I do not mean to say anything at all about the bodily processes which occur in the eye and the optic nerves and the brain. I have no doubt, myself, that these bodily processes *do* occur, when we see; and that physiologists really do *know* a great deal about them. But all that I shall mean by "*seeing*," and all that I wish to talk about, is the mental occurrence—the act of consciousness—which occurs (as is supposed) as a consequence of or accompaniment of these bodily processes . . . . It is solely with *seeing*, in this sense— seeing as an act of consciousness which we can all of us directly observe as happening in our own minds—that I am now concerned (p. 29).

Ryle, of course, would not accept this dichotomy of the mental and the physical, nor would he accept Moore's introspective approach. Nevertheless, he too rejects the relevance of physical and physiological explanations in a discussion of what is involved in perceiving. In addition, he holds that the science of psychology is only capable of explaining *mistakes* in perception, and that it has nothing to tell us concerning veridical perception.[6] Having dealt elsewhere with

are ideas in Locke's sense " (Eleventh Edition [1908], p. 548; also to be found in Ward, *Psychological Principles*, p. 27). Stout takes a not dissimilar view in the two books which Moore cited: cf. *Analytic Psychology*, I, 3–8, 15, 19–21, 26–35, and *A Manual of Psychology*, p. 7 and pp. 46–56.

On the other hand, in his *Principles of Psychology* James explicitly rejected a mentalistic definition of the province of psychology, and argued against the sort of separation of mental and physical occurrences which characterized Moore's treatment of perceiving. He defined psychology as "the Science of Mental Life, both of its phenomena and their conditions" (I, 1) and argued that "bodily experiences . . . and more particularly brain experiences, must take a place amongst those conditions of the mental life of which Psychology need take account" (*ibid.*, p. 4).

[6] D. W. Hamlyn has attempted to support Ryle's contention and to work it out in detail (cf. *The Psychology of Perception*). He says: "The ways in which we perceive something can be divided into two classes—the right ways and the wrong ways. Indeed, one way of perceiving something—the right

this contention, I shall not return to it here.[7] I have only noted his views and those of Moore with respect to psychology in order to illustrate the fact that any arguments against direct realism which are based on a causal analysis of sense perception would not be regarded by them as directly relevant to their positions. Consequently, I shall not at this point criticize direct realism by means of the sorts of argument brought forward by Strang, even though it is with such arguments that I have the strongest sympathy.

The approach which I shall instead adopt is one that takes its point of departure not from the sciences but from a consideration of ordinary experience. I shall attempt to show that if one chooses examples of sense perception which are different from the types of examples actually chosen by Moore and by Ryle, it is not in fact plausible to hold that what is directly present to us is an object whose characteristics are precisely what we perceive them to be. What will become apparent is the need to distinguish percept from object, and when this distinction is drawn the question can arise as to how the qualities of each are to be related to the qualities possessed by the other. To establish this point through an appeal to cases which are familiar in ordinary experience, I shall first consider the case in which what we see is not an object such as a book, a chair, or a tree, but the sun or a star. It will then be useful to compare such a case with hearing a sound,[8] and then with other

---

way—may be distinguished from all others" (p. 11). He then draws the following conclusion:

> It is clear that there may be laws, or at any rate generalizations, applicable to the generation of illusions . . . . But what could be said in a general way about correct perceivings? As I have attempted to make clear, we can say that people see things correctly under normal conditions, but what conditions are normal can be determined only negatively by contrast with abnormal conditions which produce illusions (p. 16).

It is amusing to note that this view, which is by no means new, was characteristic of those nineteenth-century psychologists who invoked "faculties of the mind" as the basis for psychological understanding (cf. James, *Principles of Psychology*, I, 2 f.).

[7] Cf. "Professor Ryle and Psychology."

[8] It is worth noting that Moore assumed that whatever might be established regarding sight would be "easily transferable, *mutatis mutandis*, to all the other senses by which we can be said to perceive material objects" (*Some*

cases drawn from our visual experience. Through these examples it
will be possible to suggest that not everything about perceiving is as
simple and as obvious as Moore and Ryle would have us believe.[9]
When we say that we see the sun or a star we presumably do not
believe that the sun which we see is actually to be described as a
dazzlingly bright disk, nor that the star itself is a tiny glittering
point of bluish light: in our everyday world we accept sun and stars
as being immensely distant objects which possess characteristics quite
different from those which we would attribute to them if we were
merely describing what is visible to us when we look at the heavens.
Nonetheless, when we say that we see a particular star we mean
that what we see *is* a star. These two beliefs—that we do really see
a star, and yet that the star does not possess the properties which
we see it as having—are not inconsistent. What saves them from
inconsistency is the fact that in such cases we readily interpret the
relationship between the object which we see and the qualities which
that object appears to us to possess as a causal relationship. Such a
causal relationship need not be clearly conceived, and on the level
of common sense it presumably is not: it merely involves a belief
that if there were no object of the sort that we refer to as "a star"

Main Problems of Philosophy, p. 29). For a brief comparison of sounds and
visual sense data, cf. Moore's *Commonplace Book 1919–1953*, p. 49.

[9] To be sure, there is one passage in an early essay, "The Nature and
Reality of Objects of Perception," in which Moore discusses two types of case
of visual perception which appear to raise difficulties for his direct realism:
(a) the size which a distant object, such as the moon, appears to us to have,
and (b) differences in the color of blood when seen by the naked eye and
when seen under a microscope (cf. *Philosophical Studies*, pp. 93–95). However, the context in which he discussed these cases was in terms of whether
they lent support to Berkeley's position: he did not work out in any detail what
implications they might have had for his own position. A similar limitation is
to be found in an earlier passage in the same essay, in which he refers to the
fact that light waves of varying frequencies are associated with our perception of
colors (pp. 89–90): this fact does not then raise for him the problem of
primary and secondary qualities, for the context in which he posed it was
solely that of whether Berkeley was correct in holding that "exists" means
"is perceived" (cf. p. 91).
Because of the restricted uses to which Moore put these illustrations, I do
not believe that what I shall later have to say concerning his choice of
examples and his neglect of scientific entities is misleading.

we should not be presented with the particular qualities by means of which we describe how the star looks to us.[10] To separate the look of a star—its appearance to us—and the star itself, is not to invent entities for philosophic purposes. To speak of the faint and shimmering light of a star when we look at the heavens is a perfectly natural mode of discourse, and one which we take to be descriptive of our experience. The point of light which we see is a token of the star's presence, as a faint and flashing light may be the token of an airplane's presence. In neither case need we think—nor need we speak—as if what were there for us to see was to be identified with the object itself.

However, there must be acknowledged to be a very great difference between what we take for granted about seeing a star and what we take for granted when, for example, we see a book. In the latter case we would assuredly reject the view that what we are actually seeing is *caused* by the object at which we are looking; rather, what we see strikes us as *being* the object. And this is true even in those cases in which we feel that there is something deceptive about the appearance of what we see. For example, if we regard the illumination as responsible for making a book appear a different color from the color which we would say that it actually is, we would nonetheless hold that what we have immediately before us is the book

---

[10] In *Perception, Physics, and Reality*, C. D. Broad showed how the difficulties of naïve realism tend to push common sense toward the acceptance of a causal theory of perception (for his summary statement, cf. pp. 186–87). I have no doubt that this is true. It is also true that an acceptance of scientific astronomy would lead us to distinguish, as I have distinguished, between the star's nature and the appearance which it presents to us. However, in addition, I should like to point out that there may well be phenomenological grounds on which—without argument—we distinguish between what we call " things " and what we regard as " mere appearances," and that stars (as we see them) may possess those phenomenological characteristics which lead us to classify them as belonging to the latter rather than to the former group. Among such phenomenological characteristics I might mention indefiniteness of contour, lack of fixity (e. g., flickering), absence of perceived microstructure, and color which is not surface-color. How these various characteristics are interrelated is, of course, an interesting phenomenological and psychological problem. Even more important, however, is the question of whether our discrimination between appearances and objects in terms of these characteristics (and their opposites) is wholly due to past experience.

itself, and not some token of it. On the other hand, when we look at a star we are willing to accept the view that what we see is in some sense an image of the star, for if we were to *describe* (and not merely name) what we are at present seeing, we would not regard that description as an accurate description of the star itself: what we see (we might say) is the light of the star. It is because of this difference between the two cases that direct realists take the case of seeing a book as paradigmatic for an analysis of perception, while one who holds a representative theory of perception would wish to devote careful attention to cases such as those of a star, in which distance substantially alters the appearance of objects.[11]

Now, if we consider audition as well as considering vision we will, I submit, be inclined to place an added emphasis on the case of seeing the star, for in our auditory experience there are more parallels to it than there are to what we regard as given when we see a book. For example, when we hear the sound of a bell, we do not assume that the sound which we hear is a property of the bell itself, but, rather, that it is *caused* by the bell. To be sure, we may identify

---

[11] It is, of course, not only the phenomenological difference in the two cases which is of particular interest to those who hold a representative theory of perception; there is also involved the inference which is to be drawn from the finite velocity of light. As A. O. Lovejoy put the matter:

Roemer's observation in 1675 . . . was as significant for epistemology as it was for physics and astronomy. It appeared definitely to forbid that naïvely realistic way of taking the content of visual perception to which all men at first naturally incline. The doctrine of the finite velocity of light meant that the sense from which most of our information about the world beyond our epidermal surfaces is derived never discloses anything which (in Francis Bacon's phrase) "really exists" in that world, at the instant at which it indubitably exists in perception (*The Revolt Against Dualism*, p. 19).

In G. E. Moore's *Commonplace Book 1919–1953* there is an entry (p. 219) which is entitled "Velocity of Light Argument." In that entry Moore refers to "L," whom the editor tentatively identifies as Lazerowitz. It is conceivable that the reference was to Lovejoy, for the entry seems to have dated from the period when Moore was resident in Swarthmore (cf. references to Blanshard and Köhler in surrounding passages), and Lovejoy gave a series of lectures at Swarthmore College at that time. On at least one occasion Lovejoy participated with Moore in a faculty discussion of philosophic problems, but I do not recall that this particular topic came up for discussion. Moore's entry does not, however, touch the point which is at issue in my discussion, above.

the tone of the sound with one particular bell and be able to identify this bell by *its* sound. However, when we say that we hear the bell, we are willing to acknowledge that we are speaking elliptically: that it is the *sound* of the bell which we are hearing. And this, then, is like the case of the star, in which, as we have noted, we would be not unwilling to say that what we see is the light of the star. In this case too we might, when challenged, say that when we claim to see the star we are speaking elliptically.

To be sure, there are significant differences between all visual experiences and all auditory experiences. One of these differences is that (in general) we regard the colors, the shapes, and the other visual properties of objects as "belonging" to these objects; that is to say, we regard them as qualities which are always present when the object is present. On the other hand, we do not so regard the sound of a bell. For example, while we expect a bell always to make a certain sort of sound when it is struck, we do not expect it always to be sounding; on the other hand, we do expect it always to be characterized by a particular color and shape. And, in this particular respect, seeing a star is more like other visual experiences than it is like our experience of sounds: we expect a particular star to display the same visual properties from one night to the next, so long as our vision of it is not impeded. On the basis of this difference between the two sorts of case, the direct realist might contend that an analysis of audition is not really germane to an analysis of vision; he might then go on to hold that since it is presumably not by means of hearing, but by means of sight and touch, that we come to believe in an external world,[12] whatever may be the case with respect to audition is of limited significance for the fundamental questions of epistemology.

Against such a counterclaim I should like to point out that there are a number of similarities between audition and vision which everyone accepts, and that it is by no means certain that these similarities are epistemologically irrelevant. For example, both seeing and hearing involve the use of specific sense organs. Furthermore, if

[12] This, for example, is the position adopted by H. H. Price (cf. *Perception,* p. 2).

we wish to remain in conformity with ordinary modes of thought we may put the matter much more strongly: not only are our eyes and our ears in some sense "involved" in seeing and in hearing, but we see and we hear because of the way in which these parts of our bodies are affected by external objects. Another similarity is to be found in the fact that we recognize that the medium between external objects and our sense organs will affect our vision and our hearing in a number of analogous ways. For example, the distance of the object from us will readily be discovered to affect how well we see and how well we hear, and it will have such effects in relatively consistent ways in each of these sense modalities. Furthermore, in both cases we can discover how the nature of the medium (i. e., what exists between us and the object), or what changes occur in it, may affect what is seen or heard. In both sense modalities, too, we find that there are what we term illusions; cases of hallucinations are also to be found in both. Taking these similarities into account (and no effort has here been made to mention all that might be found), we may regard it as likely that an epistemological discussion of our visual experience will be helped by a comparison with what occurs in hearing, even though it is true that, in most cases, there is a significant difference between the way in which we relate a sound to its cause and the way in which we ascribe a visual quality to an object.[13]

Once, then, it is suggested to us by our ordinary experience (and quite independently of the experimental sciences) that there may be an analogy between what occurs in hearing and in vision, the case of the star should take on added significance. For it should occur to us to wonder whether our usual identification of *what* we see with the object *which* we see is not a mistake. Such an identification is not present in hearing; nor, as we have noted, is it present in the case of seeing a star. To be sure, one might wish to

[13] However, it should not be assumed that all which we see is actually attributed to objects as comprising a quality of such objects. And if we wished to press these cases and demonstrate their importance for our ordinary perception of material objects, the apparent differences between the world of vision and the world of sound would break down even farther. Among the phenomena to which I here have reference are film colors; also some volume colors; and, more especially, shadows and highlights of illumination.

deny that what we believe with respect to seeing an object which is as distant as a star should be taken as having a bearing upon what we may legitimately take to be true with respect to other cases of vision. However, we are not tempted to hold that what occurs when we look at an object which is fifty yards away is different from what occurs when we look at something closer at hand; nor do we introduce a different sort of analysis of vision when we extend this distance from fifty yards to a mile, or to five miles. And, so far as I know, no one has ever claimed that a different analysis of vision should be proposed in the case of seeing the moon, the sun, or a star, than is proposed when we look at objects in our immediate environment. In brief, there seems to be a continuity running through all of these cases. Under these circumstances, and bearing in mind the analogy of hearing, it is not surprising that in the past, even before the rise of the modern experimental sciences, many people should have felt it necessary to say that even in those cases in which we regard our vision as being most trustworthy, we have no right to assume that when we describe the way that an object looks to us, we are giving a description which may also be said to describe the properties which that object possesses independently of us.

Perhaps direct realists of the stamp of Ryle, or of Moore, could find some way of analyzing the perceptual experience of seeing a star which would be consistent both with our ordinary beliefs about the nature of stars and with direct realism. While no such analysis has occurred to me, the fact that they do not discuss so obvious a problem leads me not to wish to overemphasize it. Therefore, I shall now abandon this illustration and take up once again the three cases of visual perception which I previously discussed in connection with Hume: the oar which looks bent when it is immersed in water, but is straight; the tower which looks round when seen from a distance, but turns out to be square; and the changes in coloration which we observe as we approach a distant mountain.[14] I shall argue that even without appealing to the experimental findings of physics and physiology, there is good reason in all three of these cases, no

[14] Cf. pp. 125–34, above.

less than in the case of the star, to hold that what is immediately presented to us in vision is not to be taken as identical with the properties of the object which we would describe ourselves as seeing, but is to be regarded as an effect of the action of that object upon us.

It will be recalled that in my previous discussion of these cases I attempted to show that the contradictions which we find in our perceptual experience do not justify an acceptance of subjectivism. In the course of my analysis of each case I also pointed out that not every *difference* in our perceptual experience on two different occasions, nor even every difference in what one perceives by means of different sense modalities, actually involves a contradiction. That there sometimes are contradictions in our perceptual experience I should not for a moment wish to deny; however, they are not my present concern. For my present purposes it will be sufficient to discuss those cases in which there simply are *differences* between what we perceive on different occasions or by means of different sense modalities, since what holds in such cases will hold *a fortiori* where these differences are in fact contradictory.

Let us first take the case of the mountains which look blue from afar, but which, as we approach them, appear to be a motley array of colors. In such a case, as I have pointed out, we are not likely to feel that the different views of the mountain are in any sense in conflict with one another, for it is unlikely that, under the circumstances, we are apt to think that the mountains should be described as having any one color. (Mountains, we should be more inclined to hold, take on different colors as the light upon them changes with the time of day, with passing clouds, and also with the nearness or remoteness of our view of them.)[15] This signifies, however, that though we are not for a moment denying that the mountains exist independently of us, and though we are not saying that they are in any sense colorless, we are saying that there is no one color (or set of colors) which is *their* color. What we would perfectly normally say is that they appear to be of different colors at noon, at sunset, in the moonlight, etc., just as they do at various seasons. In a

---

[15] Cf. J. L. Austin on the question of what is " the *real* shape " of a cat, etc., in *Sense and Sensibilia*, pp. 65–67.

similar manner (though with a difference which we shall note), we do not find it contradictory that, from different angles of vision, the shape of the mountain appears to change. We do not regard these different spatial forms as contradictory, for we have learned to expect that the appearance of objects such as mountains will continuously shift as we move with respect to them.[16] In the case of shape as well as in the case of color we would then find it perfectly acceptable to speak of "the appearance of the mountain"—although in speaking in this way we would certainly not be meaning to say that the mountain is "an appearance," and not "a reality"; nor would we thereby be implying that the mountain is in fact nothing else than the sum of its appearances.

However, even though the case of the shape of the mountain and its color are analogous in the one respect which I have here pointed out, there is at least one other respect in which we would insist that there is a significant difference between them. In the case of shape, we would reject the view that the shape of the mountain is to be regarded as being merely the particular shape which, from any one point of view, we see it to be. Nor would we hold that its shape is simply the sum of the different ways in which its contours appear to us when we see it from different angles—though this might constitute the only means which we have of plotting its shape. However, it would not in the least contradict our ordinary beliefs were we to say that there is no such thing as *the* color of the mountain: the mountain, we may well say, has whatever color it assumes at any time, or from any distance. And if we were in fact to speak in the latter way, we should of course expect it to be understood that we were speaking of the color which it takes on

[16] However, some of the specific changes which we may see from different angles may be quite unexpected.

I should also like to point out that what is true with respect to an object such as a mountain is not necessarily true of all objects. For example, objects which are much smaller in size, and which are of what appears to be a symmetrical shape, may have a greater "shape-constancy" in their specific appearances, and generally present fewer surprises of the sort here mentioned. Traditional epistemological discussions of the supposed elliptical shape of the penny have not usually taken the fact of "shape-constancy" (on the analogy of "color-constancy," and "brightness-constancy") into account.

for an observer under specific conditions, since we would certainly not wish to suggest that a mountain assumes a color in the sense in which a character actor assumes a role. It is we, as observers, who see it as now having one color and now another, depending in part at least on the conditions under which we see it.

I do not wish to insist that the foregoing description is impeccable from the point of view of common-sense notions and of ordinary language.[17] I have merely sought to suggest that in this particular case it would be implausible to hold that, if we have not made a mistake, what we see when we look at an object should always be held to be a property of that object, and not some particular "look" which it has. And my description should have served to suggest that we are willing to acknowledge that, in some cases at least, some of the characteristics of the objects with which we are visually acquainted may depend upon relations which are extrinsic to the natures of these objects, for example, on the light which falls upon them, or the position from which we happen to see them. In short, on perfectly straightforward phenomenological grounds, and without any appeal to physical or physiological explanations of what is involved in vision, we find through comparisons of what we see on different occasions that the visual aspects of objects may not, in some cases, be identified with the object which we regard as being independent of us and which we would say that we were actually seeing. To be sure, I have not claimed that this is in the same sense true in all cases. For example, in the case of shape, we might insist that— if we had not made a mistake—each of the visual aspects of the contour of an object is an aspect of its actual shape. My point has only been that we do not necessarily regard this to be true in the case of colors. In the sort of case which has here been under discussion (and there are other sorts of cases which might also have been considered, as the phenomenon of iridescence will serve to remind us), we do not in fact take the colors which we see to be

---

[17] However, my use of the verb "to assume" should not, I believe, be regarded as strained. The *Oxford English Dictionary* does in fact list one not inapposite quotation from a book on dyes: "Mercury with a larger quantity of oxygen assumes a red color" (cf. sense II, 4 under "Assume").

"in" the mountain, but to be the ways in which the mountain appears to us.[18]

The same general conclusion follows from the other two cases, which I shall be able to treat more briefly. Consider first the case of the oar which looks bent when immersed in the water. The problem which is of present concern to me is not a question of how we eliminate contradictions in sense perception and how we in fact decide that the oar is actually straight; with some aspects of that question I have already dealt. Rather, what I here wish to point out is simply what can be learned about the *nature* of sense perception and not about its reliability, from a case such as this. Let us then assume that we can be said to know that the oar is straight, and that the fact that it looks bent is recognized by us to be some sort of illusion. Nevertheless, we still see it as being bent at an angle where the water line cuts across it, and as we raise and lower the oar we will see this angle moving up and down along its length. Now, the question which I wish to raise may be put in a traditional way by asking: "*Where* does this illusory appearance exist?" The common-sense answer to that question—which seems to me perfectly satisfactory, and in the end inescapable—is that it exists where we see it. Where we see the bent oar is, of course, in the water, just as when we look into a mirror we may be said to be seeing our face *in* the mirror. However, there are oddities about the sense in which the bent oar is "in" the water. In the first place it is obvious that we do not regard what we see as being a property of the water, any more than we would describe the face which we see in the mirror as

---

[18] It is to be noted that I here use the first person plural, and that nothing which I have said has been phrased in a way which would suggest that what appears to the observer is necessarily "private," in one accepted meaning of that term. For example, what I have said in this context would be wholly compatible with holding that even the most fleeting color which appears on the face of the mountain fulfils what Ryle would use as a criterion for a public fact: I would expect anyone else "who was in a condition and position to see [it] properly" to identify this color with the color of a common set of objects (cf. *The Concept of Mind*, p. 220; also p. 202). However, such a designation of what is "public" and what is "private" has little to do with what those who have held a representative theory of perception have usually meant by the privacy of our sensations.

in any sense belonging to the mirror. Furthermore (and perhaps not unconnected with this fact), we may note that even though the term "in" carries a specifically spatial connotation when we say that we see the bent oar in the water, or the face in the mirror, so long as we recognize that we are dealing with an illusion (or, alternatively, with a mirror image), the space in which we see the oar, or the face, to be located is not seen as continuous with the three-dimensional space in which our bodies and other material objects are experienced as existing. Thus, to speak accurately we should not say that the bent oar which we see is really the oar: to speak precisely, we should say that we do not actually see the oar below the water line, but we see an image of it. And this, I submit, is what we may also properly say when we see ourselves in a mirror: we do not actually see our face, but an image of our face. That we are not usually inclined to speak in this way is due to the fact that we take the properties of that image to be similar to our actual features, and thus are inclined to identify them. In the case of the oar which appears bent we do not make such an identification, and we are therefore inclined to speak of a reflection of the oar, the oar's image, or the look which the oar has.

Now, the importance of this case in the context of my present discussion is that it points to the same sort of fact with which we became acquainted when we discussed seeing a star: that what is presented to us is frequently not taken by us to be the object itself, but is regarded as an effect produced by that object, or a token of it. And in the case of the bent oar, such a distinction is not dependent upon any knowledge derived from the physical sciences. To be sure, it might be thought that this case should not be taken as significant for a general theory of sense perception since it involves what is to be classified as a case of an optical illusion. However, there are other cases which do not, in a strict sense, involve what we should be apt to classify as illusions, but in which we distinguish between the object itself and the appearance which it has for us. For example, when on a foggy night we see the usually distinct street lights as softened in brilliance and contour, it is as if they were pendulous, glowing objects whose shape and color we would describe in quite different adjectives from those which we would ordinarily use in describing the same lights on other occasions. Here there is no question of an illusion: the lights simply look different to us on this occasion. To be sure, if we wish to do so, we can speak of the fog as cloaking or masking or hiding the qualities of the objects at which we are looking, but even if we were to say this we should be distinguishing between the appearance of the object and its characteristics, and we should be doing so in a case in which we would not normally speak of an illusion, or even of a mistake in our perceptual judgment. And for my part, I should say that a more accurate and natural way of speaking of such a case is simply to say that on different occasions the same objects may have quite different appearances. To say this, however, is to raise once again the question which has always given rise to questions regarding the nature of sense perception: namely, what relation is there between what we see, or touch, or hear, and the characteristics which material objects may be said to possess independently of the variations which we perceive them as having.

Finally, let us consider the case in which a tower appears to be round when seen from a distance, but when we approach it we see that it is square. In such a case there is, of course, a contradiction and we readily admit that in our more distant view of it we were mistaken. We would then be likely to say that it merely *appeared* round to us. In such a use of "appear" there is, of course, the notion of "seemed-to-be-but-was-not." However, there is also another sense of "appear" which does not carry that connotation: [19] we may, for example, say "I thought the tower was round, but it now appears to be square." In other words, there is a perfectly conventional sense of "appears" in which its meaning is "looks." Now, in the case of the tower, as in the other two cases which we have been discussing, a single material object may look quite different under different conditions, and the thesis of direct realism therefore seems to me to demand a more sophisticated and elaborate defense than it has been given by, say, Ryle or Moore. However, there is a further point which emerges with respect to the case of the tower which, I believe, counts heavily against Ryle's general position.

[19] In the *Oxford English Dictionary* compare sense 3 with sense 11 under "Appear," and sense 11 with sense 12 under "Appearance."

Ryle's direct realism, it seems to me, confuses what might be called "the classificatory use" of a descriptive adjective with those cases in which we use words in an attempt to give a careful description of a material object. For example, in our ordinary experience we are apt to classify all towers as being either round or square, and whatever finer distinctions might be drawn are, in general, of little concern to us. Thus, we are perfectly willing to say of a tower that it is a square tower, even if a carpenter or a mason points out to us that no two of its sides are exactly parallel, and that no two of them are exactly equal in length. And as we approach more closely to the tower we too will notice this fact. However, for many of our ordinary purposes the rough and ready classificatory description of the tower as "square" is all that we are concerned with, and if such is the case we are not likely to dwell on the differences between the look of the tower from the middle distance and how it appears to us when we stand near its base: in both cases we may be content to refer to it as a square tower, since it is a not-round tower. Therefore, so long as one is only interested in how people classify objects by means of the adjectives which they use, Ryle's form of direct realism may seem sensible and plausible. However, in the history of the theory of sense perception a quite different problem was raised: it was asked how the look of objects is related to the nature of these objects. Ryle's direct realism fails to provide us with any answer to *that* question, or if it does provide such an answer it will not be an answer which is likely to prove acceptable to any carpenter or mason who finds that he needs a plumb line, a level, and a rule.

That Ryle fails to deal with most of the traditional problems of perception may be made even more clear if we consider the highly restricted ways in which he uses the term "perceive," and its cognates such as "see" or "hear." In *The Concept of Mind* one may note that these terms were primarily used as equivalent to "identifying," "noticing," and "recognizing," and that in the index of that work the entry under "Perception" refers one to "Observation." With respect to our use of the latter term, Ryle says:

We use the verb "to observe" in two ways. In one use, to say that someone is observing something is to say that he is trying, with or without success, to find out something about it by doing at least some looking, listening, savouring, smelling, or feeling. In another use a person is said to have observed something, when his exploration has been successful, i. e. that he has found something out by some such methods.[20]

I suppose that the behavioristic bias of this characterization of what is involved in perception is clear. What must be explicitly noted, however, is the way in which Ryle's use of the verbs "looking," "listening," etc., permits him to glide over the problem of *what* we see when we look, or listen, taste, smell, or touch. To be sure, if it is a bird which we are observing what we see is the bird, and if it is a print at which we are looking, what we see is that print; however, in both cases we may have to attend to the particular qualitative appearance of the object if we are to identify the bird as a downy woodpecker rather than a flicker, or the print as a reproduction of a lithograph rather than the lithograph itself. And the same may be said of the taste of the wine which we savor, the odors of the cooking which we smell, and the textures of the objects which we feel. To be sure, our daily life includes many cases in which we would accept the way in which Ryle characterizes perceiving,[21] but there are other cases in which he totally fails to make clear what it means to look at, observe, inspect, contemplate, become cognizant of, or see, hear, taste, and feel. In thus leaving out of account the *qualia* which are presented to us in our experience of objects, his discussion conceals the indisputable fact that *what* we see or hear when we look or listen depends on the nature of our sense organs, upon the objects which affect those organs, and in what ways they do so. It was the recognition of such influences which has in the past given rise to philosophic problems regarding sense perception.[22]

---

[20] *The Concept of Mind*, p. 222.

[21] Even in those cases in which we would not be inclined to reject his characterization of perceiving, that characterization does not constitute a very probing analysis. For example, it fails to suggest anything about the ways in which the specific qualities of objects influence their recognizability, or to what extent perceived similarities affect the ways in which we classify objects, or in what ways qualities help to mark objects as possessing an enduring identity.

[22] It appears to me that the issues which are raised by illusions and by

Had Ryle not contented himself with offering a refutation of the subjectivist principle, and had he not also confined himself to too narrow a range of perceptual experiences, he would doubtless have seen that these problems are not in fact avoidable, and that they inevitably lead us to give a scientific as well as a phenomenological account of what perception involves.[23]

II

The same conclusion may be approached from the opposite direction. Whereas my concern in the preceding section was to show that an examination of directly accessible facts concerning perceptual experience serves to undercut the sort of direct realism which characterizes the thought of Ryle or of Moore, I shall now attempt to establish that their claims regarding our knowledge of material objects are incompatible with what is to be learned about the nature of these objects from the best attested results of the physical sciences. Thus, their direct realism will have been attacked both from the point of view of what it implies regarding human perception and what it implies concerning the nature of the physical world.

As we have noted, Moore confined himself to a discussion of cases in which what we perceive are familiar molar objects; typical of the objects which he mentioned were his own body, an envelope, a coin, a door, an inkstand, a sofa, and a tree. Furthermore, he did not apparently regard it as necessary to discuss what specific qualities such objects possessed; for the purpose of refuting idealism he merely examined the meanings of the general assertions which ideal-

erroneous perceptual judgments may be subsumed under the above problems. If not, I would nonetheless insist that at least *one* fundamental source for the traditional philosophic problems regarding sense perception is to be found in the causal knowledge which we possess regarding vision, hearing, touch, and the other sense modalities.

[23] In the end, Ryle admits that "there are all sorts of important connexions between the things that we all know, and have to know, about seeing and hearing and the things which have been and will be discovered in the sciences of optics, acoustics, neurophysiology and the rest" (*Dilemmas*, p. 110). However, just what some of these connections may be, and what their implications are, is left unexplored.

ists had made. As a consequence of this mode of argumentation, problems such as those concerning illusions or the distinction between the so-called primary and secondary qualities of objects did not play a significant role in his epistemology. Whatever may be the advantages of this mode of argumentation, it has the overwhelming disadvantage of failing to make clear the scope of what it actually establishes. In the first place, even if Moore's method could establish the independent existence of a particular object, such as an inkstand, it does not follow from any of Moore's arguments that the perceived characteristics of that object are characteristics which exist in it independently of its relations to us as percipients, and independently of its relations to other things. For example, as Berkeley insisted, no epistemological idealist need deny our common-sense descriptions of objects: an idealist, no less than a realist, may speak of a particular inkstand as being made of transparent glass which is slightly greenish in tinge, as smooth to the touch, as capable of holding ink, and he may say that it weighs six ounces, etc. Therefore, our ability to give such descriptions does not in the least settle the question of what properties, if any, exist in this particular object independently of its relations to other objects in the physical environment (such as light, or the gravitational field of the earth), or independently of its relations to us. It is when questions of this last type are raised that problems concerning idealism, phenomenalism, and realism inevitably arise; and it is precisely here that we find a lack of specificity in Moore's position. Having sought to disprove the statements made by some epistemological idealists, he failed to examine in any detail what the position of Common Sense, upon which he relied, was itself willing to affirm. Under these circumstances it is not surprising that he saw no problem with respect to whether such affirmations were warranted, nor that he failed to examine how, if at all, they were to be reconciled with the descriptions of material objects which physical scientists give.

In the second place we may note that Moore's mode of argumentation overlooked the fact that it is by no means easy to specify what it means to say of anything that it is "a material object." [24] This

[24] I have found only one attempt in any of Moore's writings to deal with the

being so, his failure to speak of the specific qualities of the kinds of objects with which he was concerned makes it difficult to know how far his arguments are to be extended, and whether they can prove the existence of entities which differ widely from those which he named. For example, we may ask whether those arguments would apply in precisely the same ways, and with equal force, to the existence of rainbows as material objects, and whether they would serve to establish that what we perceive when we see a flash of heat lightning is something which does in fact exist in the sky as we perceive it. Such questions are not raised by Moore. As a consequence, the pitifully small list of tenets which he attributes to Common Sense fails to provide us with any adequate clues as to what general sorts of entities or events are regarded by him as existing independently of our perception of them. Had he been more specific, difficulties might have arisen; nonetheless, the implications of his position would have been clearer and one would not then have been reduced to discussing merely those sorts of familiar, middle-sized, rather homey objects which he generally chose to discuss.

And now, finally, we may note that had Moore had the slightest interest in the results of scientific inquiry he could not have used his method of simple pointing to designate what it means to say of an object that it is a material object and to prove that it exists independently of our perception of it. Actually, one looks in vain among Moore's writings for discussions of, say, molecular motions or of

---

problem of what it means to say of an entity that it is "a material object," and this attempt at a definition seems to me to have been singularly ambiguous. It is to be found in his 1910–11 lectures, which remained unpublished until 1953 (Cf. Some Main Problems of Philosophy, pp. 128–32.) According to that definition, a material object is "something which (1) does occupy space; (2) is not a sense-datum of any kind whatever and (3) is not a mind, nor an act of consciousness" (p. 131). Among the problems raised by such a definition is the fact that Moore meant by "occupying space" the same as "having a position in space" (cf. p. 128), but some entities, such as shadows, which might not be thought of as being material objects have positions in space. If they are said to be sense-data, rather than material objects, the reasons for holding this must then be made clear. In short, these criteria presupposed that everyone knew from the outset all that was needed to be known about the nature of the physical world—both what it contains and what it does not contain.

the existence of electrons, and one is at a loss to see how he would have been able either to solve or to dissolve the problems which other philosophers have raised concerning the legitimacy of regarding such objects or events as characteristic of the nature of the physical world.[25] Under these circumstances, Moore's realism avoids some of the traditional problems of epistemology at the cost of having nothing explicit to say concerning how we are to interpret what physical scientists have claimed to have been able to discover about the nature of the physical world.

In contradistinction to Moore, Gilbert Ryle has dealt directly with these problems, and has attempted to resolve them.[26] His aim, of course, has been to show that there is no occasion for us to be puzzled by questions concerning the relations between what physical scientists say about material objects and our common-sense beliefs about the nature of such objects.[27] According to him, puzzlement is banished when we recognize that the language of ordinary experience and the language of physics belong to different areas of discourse. To illustrate his view, he put forward an analogy in which he contrasted the accounts which might be given by a student and by a bookkeeper of the volumes which make up the collection of a college

[25] The sole references which I find to epistemological problems raised by the sciences have already been mentioned in note 9, above.

It is perhaps also worth noting that Moore could scarcely have held the views which he did concerning the *certainty* of human knowledge had he been deeply concerned with scientific descriptions and explanations, and had he faced the question of how these are related to the ways in which we describe and explain objects and events in ordinary life.

[26] So too did Miss L. S. Stebbing. However, the main portions of her *Philosophy and the Physicists* are directed to showing the inadequacies of the specific philosophic positions espoused by Jeans and by Eddington. In my opinion she failed to bring forward any arguments strong enough to support her conclusion that "the physical world," or "Nature," is what we are all familiar with in our ordinary daily experience, and that "the physicist's world," i. e., the world as described in the science of physics, does not exist independently of that science (cf. p. 281). Were this view true, it would not be easy to explain why the predictions of physicists are confirmed in the world which we directly observe. In failing to deal with this question, and with other similar questions, Miss Stebbing leaves her own position open to serious attack.

[27] Cf. *Dilemmas*, p. 1. Ryle deals with these problems in Chapters 1, 5, 6, and 7 of that book.

library.[28] As he pointed out, it should not occasion surprise, nor any bewilderment, to find that an accountant and a student think of books in quite different ways: for the accountant, every book will be represented by a figure indicating its cost, but no mention will be made of its content or of its scholarly value, whereas the student will be interested in the latter aspects of the books, although he may have no knowledge of, or interest in, their original monetary value. As Ryle points out, this difference between what an accountant tells us about books and what a student would tell us, should not lead us to think that different objects are being referred to by them, nor need we regard one of these ways of referring to the books as being false if the other is true. Each of the two quite different accounts is, instead, to be regarded as correct from its own point of view; what is important therefore is that we should be careful to keep our "logical geography" in order, and not confuse one account with the other. It is precisely a confusion of this sort which Ryle attributes to Eddington when Eddington, in his famous and highly rhetorical remark, spoke of there being "*two* tables" before him: the perceptible table and the table which he, as a physicist, regarded as existing independently of his perception of it.[29]

However, when one considers Ryle's use of his analogy as a means of straightening out the supposed mistakes of Eddington and of others, one finds that analogy to be seriously misleading in at least two respects. In the first place, it is unlikely that a bookkeeper and a student would be puzzled, or would ever enter into a dispute because of differences between the ways in which each would refer to or describe the books. The unlikelihood of their doing so results from the fact that no one would presumably hold that the original cost of a book (in which the accountant is interested) bears any necessary relationship to the value which that book may have for any particular student. Therefore, the two accounts are parallel accounts, and no source of conflict between them is easy to imagine. However, as every reader of Galileo, Descartes, Boyle, or Locke should know, some philosophers have claimed that the way in which

[28] *Ibid.*, pp. 75 ff.
[29] A. S. Eddington, *The Nature of the Physical World*, pp. ix ff.

one can and should explain the observed properties of material objects is in terms of the effects upon us of some of their unobserved properties. And, as the case of Eddington shows, it is precisely here, where causal relationships are in question, that the issues which Ryle sought to dissolve actually first take their rise. Therefore, to cite an analogy in which two different accounts are acknowledged to be independent of one another is not to cite an analogy which is relevant to the issue which Eddington, among others, felt himself forced to raise. The analogy would only be apt if one already knew that there is, and that there can be, no epistemologically relevant causal relationship between the characterstics which physicists attribute to material objects and the characteristics which we perceive those objects as having. However, to make this assumption is to prejudge the issue; were the case already settled, the analogy could scarcely be regarded as helpful.[30]

A second point at which one can legitimately object to Ryle's analogy lies in the fact that there is presumably no sense in which an accountant would maintain that the figures which are to be found in his ledgers are to be taken as descriptions of the books to which they refer. Therefore, one may readily hold, as Ryle in fact holds, that a particular set of symbols used by accountants relate to the books only in the sense of "applying to" or "covering" them, and

[30] In the context of another reference to the same general problem (cf. *Dilemmas*, p. 6 f.), Ryle makes a similar mistake. He assumes that nineteenth-century quarrels between science and theology arose because those who debated the issue failed to see that "geological questions could not be answered from theological premisses," and that the sorts of questions which are involved in these two areas are simply not "continuous" with one another. Thus, Ryle assumes that scientific and religious accounts of the world were simply parallel accounts, between which there should be no conflict. However, to assume that this should have been seen by the persons who were parties to the debate in the nineteenth century, and that it would have been seen by them if they had carefully analyzed the logic of their concepts, is to be guilty of an obvious anachronism. It was because of the dispute itself and because of the victory of science over the religious orthodoxy of the day that Ryle is in a position to hold that the two accounts are indeed merely parallel accounts. And, let it be noted, the decision came not through an examination of the logical geography of the concepts then in use, but by appeals to matters of fact concerning the age of the earth, the distribution of plant and animal species, the history of the Bible, etc., etc.

that they do not provide us with anything that could count as a description or a characterization of the specific objects to which they apply.[31] However, Ryle then puts forward the thesis that the way in which physical theories refer to material objects is similar to the way in which the prices in an accountant's ledgers refer to the entities which originally cost these sums: in other words, he holds that the physical sciences are not to be regarded as yielding *descriptions* of material objects, but are to be interpreted as offering accounts which merely "apply to" or "cover" such objects. This, however, is a thesis which has little to recommend it. To be sure, when physical scientists refer, say, to electrons, they are not referring to tables rather than to chairs or stones or trees; in this sense their terms do "apply to" or "cover" all material objects indifferently. However, from this it does not follow that what we in ordinary language describe as a table is not in fact composed of entities such as electrons, nor does it follow that when a physical chemist describes the molecular composition of a particular type of wood he is saying something which merely "applies to" or "covers" the table, without characterizing it. Were one to maintain such a position one would in consistency also be forced to hold that when, in ordinary speech, we say of a table that it is made of "oak" or "maple," we are also failing to characterize or to describe it, but are merely using generic terms which "apply to" or "cover" it. Furthermore, one must note that if one were to ask a physical chemist to be even more specific in his descriptions of a particular table, he might (for example) tell us a good deal about the varnish which was used on it, and he could then relate the use of this varnish to the color and sheen which we would describe the table as having. Thus physical scientists do not merely speak of material objects in general, using terms which fail to describe the nature of *specific* material objects. They can tell us a great deal about, say, any table, and can elucidate many of the ways in which it differs from other tables to which one might wish to compare it; and they do this by means of what they can tell us about entities which are not directly perceptible. Thus, while no one would take the accountant's ledgers as descriptive of the

[31] Cf. *ibid.*, p. 76, *et passim.*

characteristics possessed by those objects to which these accounts apply, there are many philosophers and scientists who do take a physicist's account of the nature of material objects as descriptive of these objects, and Ryle has not given any reasons to show why they are mistaken in doing so. Once again, then, his analogy has broken down, unless one has already accepted the epistemological conclusions which that discussion was supposed to support. And these conclusions, as I shall now attempt to show, have little plausibility when stripped of Ryle's beguiling use of analogies.

As we have noted, what Ryle wishes to hold is that the traditional dispute between direct realism and a scientific account of the nature of material objects can be avoided simply by avoiding confusions in the logical geography of our concepts. This, however, implies that there is in fact a boundary line which can be drawn between our everyday descriptions of material objects and our scientific accounts of such objects; and that the questions which arise on one side of that line are so different from the questions which arise on the other, that the concepts which are appropriate in the one case are not appropriate in the other. Thus, as we cross the boundary from common sense to science, or back again, Ryle would have us not only be aware of what country we are in, and be cognizant of what language we should speak, he would also have us believe that persons on the two sides of the boundary do in fact think in utterly different ways, and that the questions which it is polite to raise when we find ourselves in one territory become inappropriate, and are to be avoided, as soon as we cross to the other. This, however, is a conception of the relations between science and common sense for which he fails to argue, and which seems to have little or no plausibility, as the following discussion will show.

When one examines Ryle's work one can, I believe, see that his interest in the nature of material objects is confined to an interest in giving what might be termed "*functional*" descriptions of them. By a functional description I am here referring to a description of an object, such as a table, in terms of its possible uses, and in terms, also, of those of its properties, such as its style or its workmanship, which may make it of special interest to one person, or to some class of persons, rather than to others.[32] Now, it is to be noted that Descartes' interest in material objects—and the interests of those who shared his general epistemological and scientific views—were of a quite different sort. The problem in which Descartes was interested was not how we are to characterize the nature of material objects with reference to their various uses, but how we are to describe the objects and events of the material world as they exist independently of our own connections with them. For the purposes of ordinary life, Descartes was no less willing than Ryle to trust to his senses. As he said in *Meditation VI*: "I remarked in them [i. e., these bodies] hardness, heat, and all other tactile qualities, and, further, light and colour, and scents and sounds, the variety of which gave me the means of distinguishing the sky, the earth, the sea, and generally all other bodies, one from the other."[33] However, he did not conclude

[32] That Ryle is primarily interested in such descriptions, rather than in what I shall term "qualitative" descriptions, is suggested by the already noted fact that in his discussions of perception he tends to equate "perceiving" with "identifying," "recognizing," etc., and tends also to stress the classificatory nature of descriptive adjectives rather than interpreting these adjectives as referring to the specific sensible appearances (or *qualia*) of objects. (Cf. above, pp. 189–90.)

[33] Haldane and Ross translation: *The Philosophical Works of Descartes*, I, 187. In this connection, the following passages are also to be noted. First, Descartes says:

There is no doubt that in all things which nature teaches me there is some truth contained.

And, shortly thereafter, he continues:

Moreover, nature teaches me that many other bodies exist around mine, of which some are to be avoided, and others sought after. And certainly from the fact that I am sensible of different sorts of colours, sounds, scents, tastes, heat, hardness, etc., I very easily conclude that there are in the bodies from which all these diverse sense-perceptions proceed certain variations which answer to them, although possibly these are not really similar to them.

Such a dissimilarity would not, of course, entail that the perceived variations are not useful for the purposes of everyday life. Thus Descartes can say:

The nature here described truly teaches me to flee from things which cause the sensation of pain, and seek after things which communicate to me the sentiment of pleasure and so forth . . . .

And he can claim that it was for the sake of their utility that God in fact gave us the organs of sense perception which we do have. However, they

from the fact that the sensible qualities of objects were useful for the purposes of our ordinary life that these qualities provide us with an accurate guide to the nature of the independently existing physical world. Thus, Ryle's interest in how we are to describe material objects is totally different from that of others who, like Descartes, have had interests in the relations between science and sense perception.

However, it is to be noted that Ryle's concern with our functional descriptions of objects leads him to minimize and almost to overlook a fact with which his opponents have been deeply concerned: that in ordinary life we do not confine ourselves to giving functional descriptions of material objects, but we also frequently give what may be termed "qualitative" descriptions of them. In speaking of qualitative descriptions, I here wish to refer to such descriptions which one might offer of particular objects in terms of the specific sensible qualities which these objects are perceived as having; and it was with such qualitative descriptions, and their limitations, that Descartes and others have been concerned. And it has been Ryle's dominant interest in functional descriptions which has led him to overlook the need for also considering qualitative descriptions.[34] Now, when we consider these two sorts of description we do find that they are different, as is evident from the fact that a person who does not know the uses of some particular object which he sees, and thus cannot identify or recognize it, may nevertheless describe that object. And it is to be noticed that a qualitative description of an object need not be related to any specific interest in that object, apart from the

are not for that reason to be taken as reliable guides to the nature of the physical world. Thus, he concludes this section of his argument by saying:

> I have been in the habit of perverting the order of nature, because these perceptions of sense having been placed within me by nature merely for the purpose of signifying to my mind what things are beneficial or hurtful . . . . I yet avail myself of them as though they were absolute rules by which I might immediately determine the essence of the bodies which are outside me, as to which, in fact, they can teach me nothing but what is most obscure and confused. (*Ibid.*, pp. 192, 193, 194.)

[34] In the pages devoted to perception in *The Concept of Mind*, he is so anxious to make the point that while perception *does* involve having sensations, it is something more than *merely* having sensations, that he fails to examine the question of the nature of, or the need for, qualitative descriptions at all. (Cf. pp. 222–34.)

fact that the person has been asked to provide us with a faithful description of its appearance.[35] Yet, though our qualitative and our functional descriptions of an object do thus differ from one another, it is to be noted that in many cases they neatly dovetail. For example, when I describe a table I may point out certain of its specific qualities, such as its size and the smoothness of its surface, which make it particularly useful to me as a table on which to write; or I may point out that its color makes it harmonize well with the other pieces of furniture in my room. Thus, the sensible qualities of the table and the various uses of the table are not to be considered as constituting two nonhomogeneous sorts of descriptions; we pass easily back and forth between them, and both of these aspects of objects are of interest and of concern to us in ordinary life.

However, it is here to be noted that when we *describe* a material object in these common-sense ways, we do not consider ourselves to be doing something utterly different from, or incongruous with, what we are doing when we *explain* various of its qualities, or its ways of functioning. Our common-sense explanations are couched in the same language as are our common-sense descriptions. Furthermore, we are inclined to explain the ways in which objects behave in terms of the qualities which we perceive them as having. For example, we attribute the ways in which they act in varying situations to the fact that they are heavy or light, hard or soft, flexible or brittle, hot or cold, shiny or dull. In fact, in many cases the linkage between our functional descriptions of objects and our qualitative descriptions of them rests upon the fact that in ordinary experience we are apt to assume that the qualities which objects are experienced as having are sufficient to provide explanations of the uses to which they can be put.

Now, I should not expect that Ryle would object to this running

[35] While qualitative descriptions are sometimes used in laboratory experimentation concerning perception, and have sometimes been regarded as unreliable since the situation is "artificial" and "over-simplified," they are also frequent in everyday life. For example, when a person is asked to describe the appearance of another, in order that the latter may be recognized by some one who has never seen him before, the description is apt to be either primarily or purely qualitative.

together of common-sense *descriptions* of material objects and common-sense *explanations* of the ways in which these objects are observed to behave in the contexts of ordinary life. Where he wishes to draw a clear line of demarcation is not so much between description and explanation, as such, but between the territory of common sense and the territory of *scientific* explanations. However, we are now in a position to see that if there is no sharp line to be drawn between description and explanation within the territory of common sense, then such a line cannot legitimately be drawn between our common-sense explanations and our scientific explanations, in so far as the explanation of the behavior of material objects is concerned.[36]

Consider the case in which I wish to order a table and I go to a furniture store or to a furniture maker to do so. I specify that the table should be especially hard, so as not to mar under the conditions of use to which it will be put; that it should be of a certain color, to match or blend with other pieces of furniture in the room; and that it should be of a certain shape and of specific dimensions in order to be useful to me and to fit where I wish to put it. Now, the furniture maker, if he is a good craftsman, will know what wood to use, and how to design and to finish the table in order to meet these specifications; and he will know all this without recourse to any knowledge beyond that furnished by his own experience as a craftsman. Presumably, no knowledge of physics or of chemistry would be either needed or used by him in translating my specifications into a solid, well-built, and admirably functional table of just the appearance and design which I desired.

On the other hand, manufacturers are not impractical men, and experience shows that they are not acting unreasonably when they engage scientists to discover ways in which the physical properties of ordinary material objects may be altered, or when they ask the scientists whom they employ to find new, substitute materials, with superior functional properties, for the usual materials out of which

[36] Reasons might be adduced in support of the view that in so far as the behavior of human beings is concerned, a sharp line can be drawn between common-sense explanations and scientific explanations. However, with that problem I am not presently concerned.

specific types of objects are manufactured. Now, in some cases at least, the superior functional qualities which manufacturers seek are precisely the sorts of functional qualities which we value in everyday objects, such as superior hardness, or colors less subject to fading, or particular tactile qualities, or the like. Thus, although it is true that in giving their explanations of even the most common occurrences, scientists use concepts which are foreign to our ordinary speech, and which are not based upon knowledge accessible to us in our daily experience, these explanations nonetheless concern precisely the same characteristics of material objects as those with which we are deeply concerned on many occasions in our everyday experience.

If the above remarks are true, the attempt to draw a sharp line of demarcation between common-sense description and scientific explanations must fail. For we saw that there is no tendency on our part to isolate our common-sense descriptions of the nature of material objects from our common-sense explanations of how these objects will act and react in a variety of situations; and we have now further seen that we cannot wholly isolate common-sense explanation from scientific explanation. Consequently, there is no way in which we can keep our ordinary descriptions of the properties of material objects from coming into contact with the concepts which scientists use in explaining the presence, or the absence, or the changes, in these properties. In short, the boundary between what we mean when we describe a table as hard, and the concepts which scientists use to explain variations in hardness, is not a boundary which it is impossible to cross. In fact, it is a boundary which we readily cross as soon as we ask—in either the language of common sense or the language of the sciences—for explanations of what we observe. Once having crossed this boundary, it is imperative that we should know how to speak both the language of science and the language of common sense, regardless of which side of the border we regard as our spiritual home.

Now, it is undeniably true that when we speak two different languages we frequently find that what can be said directly and simply in one can only be said by indirection in the other. Therefore, we should not expect that all translations are to be performed by discovering exact equivalents for the individual terms of one language in the other. And some of the values which are to be found in one of our languages—values of precision, or of pathos—we may find to be lacking when we use the other. In this there may be cause for regret. Yet, no matter how different the capacities of the two languages may be, there is no reason to suppose that he who is bilingual must change his habits of thought, and must raise different questions about himself and his world, when he stops talking one language and starts talking the other. Were there so radical a discontinuity between science and our ordinary experience, one would not only be forced to wonder how the sciences ever took their rise; one would also be forced to wonder by what means their conclusions would ever come to be accepted as confirmed. In short, because of his concern with the niceties of the grammars of different languages, Ryle seems to have forgotten that it is people who learn to speak languages, and that the problems which interest these people do not necessarily change as they learn to speak new languages. Because so many problems remain the same, people are forced to translate from one language into the other. That some concepts in one language are not directly translatable into the other, and can only be translated by indirection, should not occasion surprise. In fact, it is precisely with the question of what are the possibilities and what are the limitations of such translations that those epistemologists with whom Ryle has so little sympathy have been concerned. To hold that their problem was a pseudo-problem, one must do more than point out that these languages are in fact different languages: one must also either show that one cannot both describe and explain in each of these languages, or one must show that such descriptions and explanations have nothing to do with one another. This, however, is not a task which Ryle has himself attempted, nor can one readily foresee success on his part were he inclined to do so.

This conclusion does not hold merely with respect to questions concerning the nature of material objects; it also holds with respect to questions concerning "perceiving." As I shall now briefly show, our common-sense characterizations of what is meant by "seeing" or "hearing" involve a number of causal assumptions, and once these assumptions are introduced as essential to an analysis of "perceiving," it is difficult to see on what grounds a philosopher could dismiss as irrelevant those more refined and complicated causal explanations which scientists give when they deal with perceptual processes.

To be sure, Ryle has claimed that verbs such as "see" or "hear" are achievement-words (like "win") and do not refer to processes: "they do not refer to anything that goes on, i. e., has a beginning, a middle and an end." [37] Let us grant that on this point Ryle is wholly correct so far as our common-sense views of seeing and hearing are concerned, since it is true that in most of our daily experience we are not cognizant of any temporal process which is involved in seeing or in hearing. Nonetheless—and this Ryle fails to note—what we ordinarily mean when we speak of perceiving something is that there is involved what might be termed "a transaction" between the perceiver and what is perceived. In other words, more than one entity is involved, and the relationship between these entities must (in some sense of "causal") be a causal relationship. Perhaps the easiest way to show that this is the case, and that when we are dealing with perception we are dealing with "a transaction," is to note how the term "perceive" (or "hear," or "see") is most frequently used. When we say that a man perceives a tree (or sees a tree), we assume not only that a tree is actually present, but also that his awareness of the tree would not be what it is if the tree were not present. Thus, if he says that he sees a tree, when no tree (or none such as he describes) stands in his line of vision, we might think that he was trying to deceive us. If we did not, we would probably assume that he was hallucinating, or that he was seeing a mirage. However, in none of these cases would we say that he *sees* (or perceives) a tree.[38] Thus, the word "see" (or "per-

---

[37] *Dilemmas*, p. 106. For his chief discussions of this point, cf. *The Concept of Mind*, pp. 149–53, and *Dilemmas*, pp. 99–109.

His insistence on the point is closely related to a fact on which we have already remarked (cf. pp. 189 f., above): that he tends to equate "perceiving" with verbs such as "noticing" and "identifying," to the neglect of other senses of "see" and "hear." However, this error is not my present concern.

[38] Whether we would in such cases say that he *sees* anything (and, if so, what it is that he sees); or whether we would insist that he does not *see* anything, but only "thinks that he sees it," is a question to which no answer

ceive"), when used in such a context, would normally be taken as entailing the fact that certain causal consequences were fulfilled: that the man who sees the tree has his eyes open, that he is looking in the direction of the tree, and that a tree is actually there. It is also to be noticed that even further causal assumptions are involved. For example, if we are to say that he sees a tree, nothing must be standing in the direct line of his vision blocking the sight of that tree, unless a mirror or other reflecting surface is present. If a mirror is in fact present we may be willing to say that he *does* see the tree, but he sees it by means of its reflection. Now, it is to be noticed that to speak in this way is to assume some knowledge of the causal conditions under which objects are reflected in mirrors: if we had no knowledge of reflected images, and of angles of reflection, we would not say that he was seeing this specific tree by means of its reflection.

That at least this minimal amount of explanatory sophistication is implicit in our use of the phrase "he sees the tree" may perhaps be made more obvious by considering the case in which we say "he hears the bell." In affirming that he does *hear* a bell which is ringing, we do not assume that the same conditions are present as we insist must be present if we are to say that he *sees* the bell. Not only do we accept it as perfectly reasonable to say that he hears the bell when the room is dark,[39] we also accept it as perfectly reasonable to say that he hears the bell even though it is in the next room. (On the other hand, we would of course not say that he *sees* the bell if the bell were in the next room and there were no aperture opening onto that room.) Thus, I submit, our ordinary, everyday use of words such as "see," "hear," "touch," or "perceive," involve causal assumptions concerning the activities of our sense organs, the rela-

---

need here be given. In fact, our ordinary language is by no means consistent in the locutions which are involved in various cases of this type.

[39] Wittgenstein, in a remark quoted by Toulmin, suggests that it would be a misuse of the word "see" for a physicist to say that he has discovered "how to see what people look like in the dark" (cf. Toulmin, *The Philosophy of Science*, p. 13 f.). While the point of the passage was a slightly different one, such a use of the word "see" constitutes a radical misuse of it only if one assumes (as we *do*) that light must be present, in order for us to see.

tions between objects and these organs, and in some cases even the sorts of physical conditions under which it is possible for objects to act through an intervening medium and affect us. If such explanatory notions enter into the ordinary meanings of perceptual terms, it seems difficult to grasp why philosophers should claim that scientific explanations, which deal with precisely the same objects and processes, are irrelevant to a discussion of what is involved in perception.[40] Thus, as in the case of our beliefs about the nature of material objects, I see no way of separating scientific questions from some of the questions which philosophers ask concerning the nature of perceiving.[41]

[40] In *Dilemmas*, Ryle uses one familiar type of argument to rebut the introduction of scientific accounts of perception. He says:

"When asked whether I do or do not see a tree, I do not dream of postponing my reply until an anatomist or physiologist has probed my insides. . . . The question of whether I have or have not seen a tree is not itself a question about the occurrence or non-occurrence of experimentally discoverable processes or states some way behind my eyelids, else no one could even make sense of the question whether he had seen a tree until he had been taught complicated lessons about what exists and occurs behind the eyelids " (p. 100 f.).

Such an argument would rebut anyone who said that the act of seeing could not occur without our understanding why it occurs; however, I know of no one who has ever made any such claim. On the other hand it fails to rebut either of the following propositions: (a) that to understand as fully as we can what occurs in perception we need to rely upon the experimental sciences, and cannot consult direct experience alone; (b) that to understand what occurs in perception is relevant to any claims which we make as to what characteristics objects possess independently of our perceiving them.

[41] It may be true (although I am not inclined to believe that it is) that the only empirical facts about perceiving which philosophers must take into account are facts which have been known for a very long time, and that current empirical investigations are of no great relevance to contemporary epistemology. This position was suggested to me in conversation by Professor Roderick Chisholm. While the particular problems which are of primary concern to him in his own epistemological writings do not perhaps demand more than this modicum of empirical interest, he has indeed left room for scientific inquiries of the sorts which I would deem likely to be fruitful. (Cf. *Perceiving: A Philosophic Study*, pp. 137 and 138, and especially Ch. X, Secs. 2 and 3.)

I find it of interest that very recently a number of other philosophic discussions of the problem of perception have been closely linked to scientific

III

In the preceding sections of the present chapter we have been concerned to criticize two widely shared views which, if accepted, would lead us to think that experimentally established facts concerning sense perception are of no direct significance to philosophical discussions of the same topic. Had we been able to draw upon the sciences of our own day, as Boyle and Locke drew upon the sciences known to them, our argument could have been more direct, and at the same time more forceful. However, since the relevance of the sciences to philosophic issues would have been challenged by those whose views we were examining, it was necessary to rely upon an approach which did not at any point depart from what may be assumed to be the common-sense position of most adult members of our society—a point of view to which those whose views we were examining make constant appeal.

Assuming the soundness of the previous arguments, we need no longer confine ourselves to the sphere of common-sense observation and explanation, but can come to grips with the problem of whether many of the investigations of physicists, physiologists, and psychologists do not in fact provide a means of solving some of the major issues which are generally conceded to be epistemological issues. There is one familiar form of argument which has frequently been used in an attempt to show that this *cannot* be the case. That argument aims to prove that an appeal to direct experience must always be granted final authority in epistemological issues, and that the sciences can never be regarded as correctives to our direct experience in matters which are of philosophic concern; it seeks to establish that position by contending that scientific inquiry ultimately involves reliance upon the inquirer's own direct experience. I shall discuss this frequently-used argument in the form in which it is to be found in H. H. Price's *Perception*:

All beliefs [about material things] are based on sight and

---

examinations of the same problems. This is obviously true of the work of J. R. Smythies, of R. J. Hirst, of Colin Strang, and of J. J. C. Smart.

touch . . . . Beliefs about imperceptibles such as molecules or electrons or X-rays are no exceptions to this. Only they are based not directly on sight and touch, but indirectly. Their direct basis consists of certain other beliefs concerning scientific instruments, photographic plates, and the like . . . . It follows that in any attempt either to analyze or to justify our beliefs concerning material things, the primary task is to consider beliefs concerning perceptible or "macroscopic" objects such as chairs and tables, cats and rocks. It follows, too, that no theory concerning "microscopic" objects can possibly be used to throw doubt upon our beliefs concerning chairs or cats or rocks, so long as these are based directly on sight and touch. Empirical Science can never be more trustworthy than perception, upon which it is based.[42]

Now, if in this passage Price were merely arguing that we cannot cast doubt on the *existence* of chairs, or cats, or rocks on the basis of scientific theories which had been established by the use of objects such as microscopes or X-ray equipment, I should not be inclined to quarrel with him. However, his position seems to be a more radical one. Taking the sentences which I have quoted in their full context, Price is holding that no theory of microscopic objects can in any way be used to challenge the descriptions of chairs, or cats, or rocks which we give on the basis of direct perceptual experience, since science itself must ultimately rest upon sense perception. This thesis seems to me to be fundamentally mistaken.

Consider Price's claim that our beliefs about imperceptible entities, such as electrons, are ultimately based upon perceptual experience. Taken in a quite general sense this is true, for it is on the basis of observed changes in the behavior of instruments under varying conditions, or on what we perceive when we are using these instruments, or on what we perceive after having used them, that we claim to find corroboration for our beliefs concerning the existence and the characteristics of entities which cannot be directly observed. However, Price's statement of his position suggests that all that is involved in building our conceptions of imperceptible entities (such as elec-

---

[42] *Perception,* p. 1.
   To cite merely one other use of this argument, cf. C. E. M. Joad "The Status of Sense Perception in Relation to Scientific Knowledge."

trons), or in verifying these conceptions, is derived from seeing or touching those instruments by means of which we measure their effects. This is surely not true. Our belief in the existence of unobservable entities normally antedates the manufacture of the instruments through which we confirm their existence, and our hypotheses concerning the properties of such entities will determine the design of the instruments which we build. Furthermore, in order to test our hypotheses concerning the unobservable entities in whose existence and nature we are interested, we must in many cases regard our instruments as possessing an unobserved microstructure of their own; we cannot attribute to them only those characteristics which are directly accessible to us through our sight and our touch. What Price overlooked is the fact that scientific *theories* always go beyond that which has been directly observed, and the verification of such theories involves more than giving a description of something that has been perceived. Whether our observations are performed with or without the aid of instruments, what is important in the verification of a theory is to relate that which we perceive to what our theory has predicted. This means, however, that we are not interested *per se* in the specific qualities of what we perceive; we are only interested in the sensible qualities of observed objects in relation to conclusions deduced from our theory. Thus, neither the fact that the ultimate basis of scientific theories is to be found in observation, nor the fact that the verification of such theories involves sense perception, entails that the directly observed characteristics of material objects provide either more information or more reliable information about the nature of those objects than can be derived from scientific theory.

If we may assume that the argument typified by our quotation from Price has now had its sting extracted, we are in a position to inquire in what ways the results of empirical scientific investigations can serve as a means of establishing epistemological conclusions; it is to this topic that the remainder of the present section will be addressed.

It would of course be a grave mistake to think that a theory of knowledge can be directly established by the methods of physics, or physiology, or psychology: scientists working in their laboratories are not engaged in epistemological research. Nor do I wish to assert that when a philosopher attempts to establish an adequate philosophic theory of human sense perception, he is attempting to do the same sort of thing which scientists attempt to do. However, the fact that the aims of scientists and philosophers are different in character does not imply that the results which are obtained in one of these fields may not be of direct significance to the results obtainable in the other. And this relationship need not be symmetric. In fact, if we examine the relations between philosophy and the sciences it is probably accurate to say that while the results of each may be relevant to the other, the ways in which they are relevant are different. Even taking psychology as an example (and the line of cleavage between psychology and philosophy is frequently less distinct than are the lines of cleavage between the sciences and philosophy in other cases), we find that the specific problems which are of concern to working scientists rarely are problems which demand a commitment to one philosophic view rather than another. Usually, it is not until a psychologist raises methodological questions concerning his discipline, or seeks to place the particular findings of this discipline within a single systematic framework, that he faces problems directly associated with philosophical issues. On the other hand, as we shall see, *before* a philosopher can begin to answer some of the questions which are of concern in almost any analysis of sense perception, he will be forced to proceed on certain assumptions regarding empirically testable matters of fact; in such cases the sciences may provide evidence which is relevant to an estimate of the truth or falsity of a specific philosophic analysis. This asymmetry in the relations between the sciences and philosophy can probably best be summarized in saying that there are some cases in which a scientific conclusion will be sufficient to refute an answer which has been proposed with respect to a philosophic question, but that a philosophic conclusion could not refute an answer proposed to a scientific question; it could only lead one to challenge the legitimacy of that question, or the import of the conclusion.

To be sure, among the statements of matters of fact which appear as basic in any philosophic discussion of sense perception there are some which are not derived from the sciences and whose truth need not be attested by any form of scientific corroboration. Such would presumably be phenomenological descriptions of our direct experience.[43] However, the epistemologist does not merely describe specific experiences of his own, he generalizes from sets of such experiences to what he believes to be true of *all* of his experience, or to what he believes to be true of the experiences of *all* men with respect to certain types of experience, or the like. It is with reference to such generalizations that scientific investigations are relevant, even though these generalizations may refer to phenomenological descriptions in which no causal propositions are asserted.[44] Take, for example,

[43] Statements concerning our use of language would also belong in this class. However, I shall not separate phenomenological and linguistic questions since I do not believe that linguistic usage can be adequately interpreted if one attempts to treat it independently of a description of our experience. However, even if my assumption in this regard should be false, I trust that the preceding sections have shown that factual statements concerning our ordinary use of language would not provide a means of either avoiding or answering the traditional questions of epistemology. If it should then be answered that these questions should not have *arisen*, and would not have arisen had not philosophers departed from ordinary linguistic usage, I would be inclined to ask for evidence that there is any language in actual use in any society in which such problems *do not* arise.

[44] Obvious generalizations of this type (and generalizations which can be shown to be erroneous under laboratory conditions, even if one should hold that they are true under the conditions of everyday life) are the following two statements made by Prichard at the outset of an essay on "Seeing Movements": "Whatever we may say, we all always in fact presuppose that movement is absolute, . . ." and, "*If* we see a *body*, we see it as from a certain point in space situated somewhere within our own body" (*Knowledge and Perception*, p. 41). Were either of these two connected generalizations to be rejected as empirically false, it seems plain (from sections 5 and 6 of the essay) that Prichard could not have used the phenomena of apparent motion to have established the point which he wished to make in the essay as a whole.

I allude to these generalizations in Prichard's essay, since in it (p. 46), and elsewhere in the volume (p. 52), he speaks in a vein contemptuous of psychologists. However, his references to psychologists suggest that he has in mind chiefly the writings of James Ward and G. F. Stout, and in particular those sections of their systematic expositions which concern nonempirical questions. Although Prichard's essays collected in this volume date from the mid-1920's and later, the work of Ward's which was cited dates from 1885, and the third edition of Stout's *Manual of Psychology* appeared in 1904. (Cf. also note 5, above, on the influence of Ward and Stout on G. E. Moore.)

Price's famous statement concerning "the given" in perceptual experience:

> When I see a tomato there is much I can doubt. I can doubt whether it is a tomato that I am seeing, and not a cleverly painted piece of wax. I can doubt whether there is any material thing there at all. Perhaps what I took for a tomato was really a reflection; perhaps I am the victim of some hallucination. One thing however I cannot doubt: that there exists a red patch of a round and somewhat bulgy shape, standing out from a background of other colour-patches, and having a certain visual depth, and that this whole field of colour is directly present to my consciousness. What the red patch is, whether a substance, or a state of a substance, or an event, whether it is physical or psychical or neither, are questions that we may doubt about. But that something is red and round then and there I cannot doubt. Whether the something persists even for a moment before and after it is present to my consciousness, whether other minds can be conscious of it as well as I, may be doubted. But that it now *exists*, and that *I* am conscious of it—by me at least who am conscious of it this cannot possibly be doubted.[45]

Though this attempts to be a straightforward and neutral description of what is given in perceptual experience, if we attempt to generalize from it all is not so clear as one might wish. For example, what Price was inspecting was something which he apprehended as an unchanging three-dimensional object, uniform in color, standing out against an unchanging background, etc. However, as those familiar with ophthalmological examinations will recognize, if it had been a point of light which Price was watching while seated in an otherwise dark room, would he have been able to distinguish the given from the not-given in terms of indubitability, and in terms of a precise "then-and-there"? And if the light had continuously dwindled in intensity, would he even have been able to identify what sense data *were* given within that which he was experiencing? Similarly, even though there are in fact many cases in which touch and hearing and smell are analogous to the seeing of the tomato, as

[45] *Perception*, p. 3.

Price claims them to be,[46] there also are cases in which what is given in these experiences does not bear the mark of indubitability: for example, we may not be sure whether or not we have felt something brush against us, or we may feel uncertain whether we detect a faint odor of gas, or we may not be able to say whether we do or do not still hear the sound of a tuning fork or the ticking of a watch as our auditory acuity is being tested. That there are these limitations inherent in Price's illustration, if it is taken to be characteristic of all sense perception, is I hope clear. To be sure, in trying to make this clear I have not appealed to the experimental findings of psychologists, but only to illustrations with which any one may be presumed to be familiar through experiences common in our society. Nonetheless, an exploration of the extent to which Price's illustration is limited, and in precisely what ways it is limited, can only be carried out with precision if experimental techniques are used. Therefore, if we are to inaugurate an epistemological discussion by raising the question of what is the nature of "the given," as Price attempts to do, the results of experiments in perception will be relevant to our discussion; they may indeed show that there is no way in which we can always connect "the given" with either "that which is here-and-now" or with that which is experienced as indubitable.[47] Yet the assumption that there are these connections is a fundamental assumption in Price's book, as it is in many other epistemological discussions of sense perception.

To be sure, the experimental investigations which would permit us to extend, or force us to limit, the objections which I have raised against Price's conception of "the given" are not investigations of the sort which he obviously had in view when he claimed that causal analyses were irrelevant to epistemology.[48] His concern was not with

[46] " Analogously, when I am in the situations called ' touching something,' ' hearing,' ' smelling it,' etc., in each case there is something which at that moment indubitably exists—a pressure . . . , a noise, a smell; and that something is directly present to my consciousness " (ibid.).

[47] Price does not distinguish between that which is experienced as indubitable and that which may be doubted on the basis of specific, assignable grounds. This seems to me to be a mistake.

[48] " Science only professes to tell us what are the causes of seeing and touch-

topics such as autokinetic movement, apparent visual speed, the relational determination of constancy phenomena, etc.; like many philosophers, the target at which he aimed was physiology, with its explanations of the causal chain of those physical and neural events which are involved in vision. However, this is too small a target if one wishes to show that empirical investigations of causal conditions are not really relevant to epistemological discussions: a great deal of the experimental work connected with perception is not physiological, but deals with the specific effects of differing objective conditions upon that of which we are immediately aware. And Price himself uses such knowledge in what he calls "the Phenomenological Argument" against naïve realism.[49] For example, he says:

> *Perspective* provides plenty of instances [of illusions]. We all know that stereoscopic vision is possible only within a relatively narrow range. Outside this range there is what is called Collapse

ing. But we want to know what seeing and touching themselves *are*. This question lies outside the sphere of Science altogether" (*Perception*, p. 2).

[49] "The Phenomenological Argument" consists of a series of specific arguments which relate to contradictions which we find in our perceptual experience. Price distinguishes between it and what he calls "the Causal Argument" against naïve realism. According to his use of these terms, the latter refers only to those causal conditions which are to be found in the nature of the medium between the percipient and the object, or in the nature and condition of the percipient's organism (and perhaps his mind). As will become clear, instances of what he calls "the Phenomenological Argument" also involve causal conditions, but these are conditions relating to the circumstances under which particular objects are perceived—apart from circumstances respecting the medium or the percipient himself. (Cf. *ibid.*, pp. 27–31.)

To be sure, when Price discusses *visual* illusions, he does not explicitly employ causal terms. The key concept in his discussion of these illusions is "being part of the surface of an object." However, he himself points out that the equivalent of this phrase when we are discussing sound or smell is "emanates from," and that this in turn involves the notion of "being caused by" (p. 28). Similarly, when he discusses the illusions of touch (p. 29 f.), causal notions are clearly involved. Under these circumstances the claim that his discussion of visual illusions involves an implicit appeal to the causal conditions under which various sense data are presented, would seem to be a warranted claim. If this be doubted, the reader need merely consider how the immediately following quotation would appear to one who accepted Berkeley's theory of the perception of distance and of the nature of causal attributions: the quotation would not then have established Price's right to speak of an illusion at all.

of Planes, and objects undergo various forms of " distortion." Thus a distant hillside which is full of protuberances, and slopes up-wards at quite a gentle angle, will appear flat and vertical, like a scene painted on cardboard. This means that the sense-datum, the colour-expanse which we sense, actually *is* flat and vertical. And if so, it cannot be part of the surface of something protuberant and gently sloping.[50]

In this account of an illusion Price is presupposing the causal influence of distance on visual experience, and he is apparently confident that there are general rules by means of which one can explain the conditions under which objects appear as three-dimensional and in perspective, and the conditions under which they do not. Thus, causal conditions are clearly relevant to Price's argument against naïve realism.

To be sure, in the above quotation, Price does not explicitly refer to any *scientific* investigations which establish and confirm the rules which he was assuming, and it would of course be mistaken to hold that a general knowledge of the effects of distance on vision first came to men through the experimental sciences. Nonetheless, after a science has developed, there is neither reason nor need for us to return to a state of prescientific innocence: theories which are experimentally established will be both more accurate and more capable of systematic amplification than the rules of common sense. It would be perverse to suppose—and I should not expect Price to suppose—that psychological investigations, and that geometrical, physical, and physiological optics tell us nothing which we do not already know about why mountains present different appearances on different occasions. To be sure, if the explanations offered by the experimental sciences did nothing more than fill in the details of what was already thoroughly familiar, their impact on our epistemo-logical views might be negligible.[51] However, there is no reason to assert that such is always the case. For example, as the science of physics has developed it has been able to discover a great deal about

---

[50] *Ibid.*, p. 28.

[51] In his reassessment of a causal theory of perception, H. P. Grice does not suggest that scientific investigations do more than this. (Cf. " The Causal Theory of Perception," pp. 143–44.)

the nature and the propagation of light, almost none of which would have been known if we had been confined to an analysis of direct experience. While this knowledge does not alter the truth that we cannot see in the dark, it cannot be regarded as merely elaborating on that fact; indeed, at crucial points it conflicts with what we take for granted in everyday life. To choose but one aspect of this example, it should be obvious that a physicist's account of the selective absorption of light by material objects demands that we abandon the notion that light merely illumines colors which are "there." As a consequence, we are forced also to alter our usual interpretation of what it means to contend that an object which is seen as blue in ordinary daylight, but which has a different color when seen under artificial illumination, "actually is blue." What we ordinarily suppose in such cases is that the object *is* blue under all circumstances, but that its blueness is masked or distorted by the unusual illumination. On the other hand, a physicist's account of the dependence of color on selective absorption is not compatible with that common-sense conviction. Since the absorption which occurs in ordinary daylight is no different in principle from that which obtains under other conditions of illumination, a physicist's explanation of why an object appears blue when seen in the daylight will be similar to his explanation of why, at other times, that object is seen as having another color: he will not say it appears blue simply because it *is* blue. Therefore, unless we were to challenge what physicists have established concerning the relation between the electromagnetic character of light and the colors which we see objects as having, we cannot interpret the phrase "actually blue" as meaning "has the characteristic 'blueness' independently of its physical relations to the light in which it is seen."

There are many other cases in which it can be shown that something which we regard as an inherent property of objects is actually dependent for its existence upon complex relations of which we are not directly aware. However, instead of turning to those cases, I should now like to show that a consideration of physiological optics permits one to make exactly the same point with respect to color vision that has just been made on the basis of physical optics.

It is universally admitted that human color blindness is associated with the structure of the eye, and it is also admitted that there are significant differences in color vision among animals and men. Philosophers have sometimes been prone to argue that these facts have no bearing upon whether colors exist in objects independently of our responses to stimuli emitted by these objects, since it might be the case that the eye is simply "an instrument" by means of which we are sensitive to what is independently there.[52] While this view might be supported as a possibility so long as one were content merely to point out that color vision is correlated with the existence of certain structures of the eye, it loses its plausibility when the causal chain involved in the physiological account of vision is examined in detail. To be satisfactory, such an account demands that color blindness be explained through showing how the cones of the retina function with respect to color vision. However, an explanation of their functioning cannot in fact be given in terms of specific *qualia*, such as red or green, existing in the object; it can only be given in terms of how the retina reacts to the physical properties of light of different frequencies. Similarly, an account of what occurs along the optic nerve, or in the visual area of the cortex, does not assume the transmission of *colors*; the terms which must be used to characterize the processes which occur in these regions are terms referring not to sensible *qualia* but to events of which we are in no case directly aware. Thus it is not in the least plausible to interpret the structures which are correlated with our ability to see red as being merely instruments by means of which redness is transmitted to us from objects which possess that quality in total independence of the nature and functioning of our organs of vision. In the whole causal chain of events which is involved in seeing an object, there is only one point at which we must appeal to the fact that the object is *red*, and that is when we are describing the color which we see it as possessing; at every other point in the explanation of these processes, contemporary physicists and physiologists dispense with the assumption that objects are to be characterized as possessing the qualities

[52] For a critical discussion of what he terms "the instrument theory," cf. Broad, *Perception, Physics, and Reality*, pp. 199 ff.

which we would ascribe to them on the basis of direct sense perception alone.

It is sometimes contended that the fact that scientists dispense with what are called "qualitative" phenomena in favor of their own forms of description of physical processes results from their bias in favor of that which is amenable to quantitative treatment.[53] However, this charge usually goes unsupported. To establish its truth one would have to show that there are other equally adequate ways of rendering the specific and detailed facts of vision intelligible to us, and that these ways involve those "qualitative" concepts which contemporary physics and physiology do not employ. This has not been shown. Therefore, one may suspect that factors other than bias account for the ways in which scientists explain what occurs in sense perception. Furthermore, the fact that physicists, physiologists, and psychologists—all of whom use somewhat different methods— have reached results which not only are consistent with each other, but are mutually reinforcing, should make us more confident that current scientific theories of sense perception represent discoveries, and are not to be regarded as the products of bias. Applied to the problems of color vision—which we have here chosen merely as one among many possible examples—these results establish the fact that

[53] This familiar view seems to me to have had its chief source in the interpretations of the scientific revolution of the seventeenth century which were brought forward by Whitehead in *Science and the Modern World* and by Burtt in *Metaphysical Foundations of Modern Physical Science*. They are not of course responsible for some of the more extreme philosophic and theological uses to which this interpretation has been put. Nonetheless, it is important to point out that the distinction between primary and secondary qualities (regardless of the terminology in which that distinction is put) can scarcely be said to have sprung from a desire to grant reality only to that which is amenable to quantitative treatment. Such a desire was not to be found in Bacon, nor was Descartes' differentiation among the qualities related to any attempt on his part to rest science on the measurable. These points seem to me to have received a misleading interpretation on the part of Whitehead in particular.

Furthermore, the qualitative-quantitative contrast is fundamentally misleading, since those features of objects which are susceptible of mensuration and of mathematical treatment are themselves qualities. A correct contrast would probably involve *both* the contrast between the sensible qualities of an object and its inferred qualities, and inherent qualities as contrasted with relationally determined qualities.

the color which we perceive an object as possessing cannot be said to exist in that object independently of (a) its inner physical structure, (b) its interaction with light of certain frequencies, (c) the presence of environing conditions, such as the background against which it is seen,[54] and (d) the physiological nature of the percipient. What is true in the case of color vision can, I believe, also be shown to apply (mutatis mutandis) in the case of all other sensible qualities. Thus, if we are to attach epistemological weight to the results of the sciences, as I have attempted to show that we must, it would seem necessary for us to establish and defend a radical critical realism.

IV

If the argument of the preceding sections of this chapter has been sound, contemporary epistemological discussions cannot afford to neglect the results of scientific inquiries concerning the nature of sense perception and concerning the nature of material objects. However, when we consider how remote from ordinary experience scientific descriptions of the world and of man have steadily become, the price which we shall have to pay if we accept the sciences as relevant to philosophic discussions will not be small: we shall apparently have to leave behind all vestiges of the realism of common sense, and place greater reliance on what we infer than on what we directly experience. Under these conditions it is not surprising that repeated attempts should be made to interpret the sciences as if they offered no means of describing what exists independently of us, but were merely ways of ordering what is presented in immediate experience. Totally apart from other grounds on which this interpretation might be criticized, its most usual formulations overlook the fact that the sciences represent an extension and refinement of methods of inquiry which we constantly use in everyday life as a means of corroborating, amending, or discounting the testimony of sense experience. As we shall see, when this fact is realized and

[54] Psychological investigations of such conditions have not been discussed above, but the facts in such cases are probably sufficiently obvious not to demand separate treatment.

its implications made clear, one of the fundamental objections to a radical critical realism will also have been overcome.

It will be recalled that, according to my use of the term, *realism* asserts that we can offer good grounds for believing that a world of physical objects exists independently of our sense perception, and that it is in fact possible to know at least some of the characteristics of these objects. At the same time I noted that a critical realist holds that at least some of the characteristics which physical objects appear to possess are not actually possessed by them. What characteristics these are, and how many such characteristics there are, is a question on which critical realists frequently disagree. The form of critical realism to which I would subscribe, and which I would designate as *a radical critical realism*, would contend that we do not have the right to identify *any* of the qualities of objects as they are directly experienced by us with the properties of objects as they exist in the physical world independently of us. While some qualities may more accurately mirror physical properties than do others, none can be assumed to be identical with what exists independently of sense perception. In short, a radical critical realism might be said to maintain a strict disjunction between the directly experienced and the independently existing. A standard objection to such a view is the claim that if we deny an identity between at least some features of that which is directly experienced and that which exists independently of experience, we would have no basis on which to assert that we have knowledge concerning the latter. However, once we recognize the continuity between the sciences and common-sense inquiries, the strength of this argument will be seen to disappear.

In order to lay a foundation capable of supporting this contention, let me remind the reader of one basic feature of the preceding essay: in it I attempted to show that the traditional types of argument against a realistic interpretation of sense experience actually presuppose a tacit acceptance of realism.[55] This was not taken to

---

[55] Since at other points I have been highly critical of Ryle's proposed way of dealing with epistemological problems, I should here like to point out that on this specific issue I am wholly in agreement with him. As he pithily

be a direct disproof of the subjectivist view of perception, for if a philosopher wished simply to postulate the truth of that view, instead of arguing for it, nothing in the preceding essay could dislodge him. However, even though philosophers may sometimes be willing to use analogous modes of simple asseveration in other instances, it is not a posture which they adopt with respect to the truth of subjectivism. The reason for this is clear: even subjectivists admit that in ordinary experience we are all realists of some sort, not subjectivists. The things which we see, hear, and touch—and to a significant degree also what we taste or smell—appear as independent of our seeing, hearing, or touching them. Were someone to ask on what this conviction is founded, we could only say that it is something to which sense experience itself seems to testify. Only when grounds are given which throw doubt on that testimony, do we really find it intelligible for anyone to ask why we should believe that what we perceive exists independently of our perceiving it. This, I take it, constitutes the element of truth in Moore's attempted proof that material bodies exist, and it also constitutes our own starting point: in direct experience we are realists, and cannot avoid being so.

However, this is only a starting point, not the fixed conclusion of an argument, for it is equally true, as the arguments against direct realism have shown, that we cannot consistently hold that everything that we experience exists precisely as we experience it. The grounds on which we base this conclusion do not necessarily derive from either scientific or philosophic considerations: in ordinary experience itself we frequently find mistakes occurring in sense perception. Nonetheless, as the preceding essay made clear, a criticism of what we perceive presupposes that there are other elements in our sense perception which we accept. Thus it is not true—as classic epistemological discussions have often seemed to suggest— that when we start to criticize the reliability of sense experience, we successively strip off quality after quality from material objects, finally leaving a bare unknown and unknowable surd. Whenever

---

remarked regarding the argument that *all* sense perception may presumably be doubted on the grounds that some specific sense presentations are acknowledged to be illusory: "A country which had no coinage would offer no scope to counterfeiters" (*Dilemmas*, p. 94).

we reject one aspect of what we were previously inclined to regard as a property of an object, it is on the basis of accepting as veridical some other aspect of our experience of such objects. For example, it will be recalled that when we rejected the view that a tower which appeared round from a distance actually was round, our rejection presupposed an acceptance of the view that an object such as a tower does not change shape as we approach it.

The stripping off of some qualities from our conceptions of the nature of independently existing objects on the basis of holding that others more truly belong to these objects, affords material for interesting and important phenomenological studies of perceptual beliefs. In general, epistemologists have failed to carry out such studies, or even to note their possible relevance to the issues with which philosophic discussions of sense perception have been concerned. For example, the standard way of classifying what we directly experience in sense perception uses two or three categories only: objects, qualities, and (in some instances) relations.[56] However, our direct experience contains more and subtler differentiations than these. For example, while it is true that one fundamental distinction which we draw is that between what we regard as independent *objects* (such as trees or chairs) and what we regard as *qualities* of these objects, we also distinguish elements within experience which we take to be dependent upon objects, but which we do not regard as qualities inherent in them. Among such elements there are not only sounds, but elements such as shadows, and the highlights which we see on the surfaces of objects. These elements belong, of course, among the immediate data of sense experience, and in most instances they are not regarded by us as being illusory: yet philosophers have not devoted any considerable amount of attention to them, being generally content to identify the immediate data of sense experience with

---

[56] The use of the term "modes," as we find it in Locke and others, does not help to expand this list of basic categories. While Locke includes among "modes" many elements of experience which are not generally classified as "qualities," his use of the term does not have a clear positive meaning. Furthermore, the particular sorts of cases which are to be cited below are not discussed by Locke in any manner which is relevant to the question of their status in perceptual experience.

what we ordinarily regard as the qualities of objects. Similarly, the illumination of a room, or the light of day and the darkness of night are neglected in most epistemological accounts of what we directly experience, for these are not elements which easily fit into a framework which distinguishes only among objects, qualities, and relations. If the effects of the rigidity of this framework on epistemological discussions should be doubted, I need perhaps only point out the perennial difficulties which philosophers have encountered in their attempts to deal with space and with time in accordance with their usual categoreal scheme.

The gradual sifting out and categorizing of the types of entities which are to be distinguished within perceptual experience is not my present concern. Nor is this the place to offer detailed criticisms of that traditional genetic account of sense perception according to which pure experience consists of *qualia* which first come to us wholly independently of one another, and which then come to be organized into objects solely in accordance with repetitive patterns and the activities of the mind. The difficulties in such a genetic account should be clear as soon as one recognizes the importance of the figure-ground contrast in our perception of objects, since the pervasive presence of that contrast is surely not explicable in terms of repetitive patterns in our past experience. Similar difficulties are to be found in the comparable genetic account of how we come to perceive relations between objects. The difficulties in these accounts will only be hidden from those who assume that whatever we are to take as original in experience must be analyzable into specific elements, each of which can be correlated with some specific stimulus on our peripheral sense organs.[57] I shall not here argue against that

---

[57] It is worth pointing out again that this assumption is commonly made. It is assuredly implicit in Locke's account of simple ideas which "enter by the senses simple and unmixed" (*Essay*, Bk. II, Ch. II, Sec. 1). I believe it is also an assumption made by Hume. However, the clearest case which can be cited is to be found in Berkeley, who claims that distance cannot be a sensible idea since it is "a line directed endwise to the eye," and is therefore only represented as a point on the retina (cf. *New Theory of Vision*, Sec. 2, and p. 141, above).

The assumption that what is "given" is to be correlated with what is reflected on our external organs of sense seems to be a common possession of

assumption, whose inadequacies should by now be familiar to all. I have been forced to bring it to the reader's attention only to make sure that it will not be a stumbling block as our analysis proceeds. The purpose of that analysis will be to show a few typical stages in the process by means of which we reject certain sensible qualities as we build up more adequate conceptions of the nature of material objects.[58]

In order to start from an instance to be found in ordinary experience rather than from one which depends upon scientific inquiry, let us take the case of the tower which looks round from a distance, but is seen to be square when viewed from closer at hand. As we have noted, our rejection of the view that the tower was actually round when it appeared so, could not be based simply on the fact that it later appeared square: this rejection also presupposed that we ascribe stability of shape to objects such as towers. "Stability," taken in this sense, would not be classified as a sensible quality by those who suppose that each quality of an object must be directly correlated with a particular stimulus affecting our peripheral sense organs. On the other hand, it cannot be regarded as a property which we *infer* on the basis of the ways in which these sensible qualities are repeatedly combined in experience: after all, objects such as towers appear to have *different* sizes and shapes when seen from different distances and angles of vision no less frequently than they appear to have the *same* size and shape when seen from the same distance and the same angle of vision. Thus, I would hold that conventional genetic theories of what we take to be the characteristic qualities of objects must be altered and enlarged in a case such as

all who attempt to account for our perception of objects as the result of the activity of the mind which unifies a congeries of *qualia* of color, shape, touch, etc., on the basis of past experience. Such an account of sense perception is not only characteristic of Locke, Berkeley, and Hume; in one form or another it is to be found in Descartes, in Mill, in Mach, and in most sense-data theorists.

[58] A close phenomenological study of this process is not to be expected here. As a matter of fact, as we shall see, such a study could only provide part of the requisite analysis of our conceptions of the physical world: in addition, one would have to take into account a major segment of the scientific development of Western thought.

this.[59] However, regardless of what may be a correct account of why this sort of stability is attributed to some material objects, this particular case has several implications which should not be overlooked. First, it shows that we regard a characteristic such as "stability" to be a fundamental property of a material object such as a tower, a property of which we are in fact more certain than we are of the color or shape which we see the tower as having at any particular moment.[60] Second, as we have noted, the basis on which we deny that one appearance of an object is veridical involves more than an acceptance of some other one of its appearances: both the denial and the acceptance are based on attributing a quality of another kind to that object. In the third place, and perhaps most importantly as a counterweight to conventional epistemological views, this particular instance shows that what we regard as a veridical description of an object may include at least some characteristics which are continuous over time, and are not analyzable without remainder into qualities which are capable of being presented to us in a series of discrete instants. Fourth, we may note that our belief

---

[59] I would suggest—without pushing the point—that among the factors leading us to predicate stability of an object with respect to the relations among its qualities is how such an object looks over short periods of time. The fact that the tower presents a steady appearance both when seen from a distance and when seen from nearby would, then, be one factor responsible for our attributing stability to it. It is to be noted that this is a very different view from that which holds that we infer stability from the results of a sum total of past sensible appearances in which both our close views and our distant views are aggregated. In the explanation suggested here, "steadiness" must be perceived in each of the two cases before the transition can occur. I do not wish to suggest, however, that this is the only factor present when we attribute stability to an object: the object-character of what is presented—or specific factors responsible for this object-character—may also play a role. (On the notion of "object-character," cf. pp. 152–54, above.)

[60] To be sure, we do not ascribe stability to all material objects in all respects. It is no less characteristic of our experience to regard some material objects as being labile in various respects. In each case we must note in what respects an object appears to have stability, or to lack it. For example, as we have noted, we do not necessarily attribute stability to the colors of a distant mountain, nor to any other cases in which the color which we see is not a surface color. As has been noted (following J. L. Austin), we do not necessarily attribute stability to the shape (or even the exact size) which an animal may appear to have. (Cf. note 15, above.)

in the stability of the relations among the various qualities of an object such as a tower will have an impact on what we expect to see when we look at it on other occasions: because we attribute stability to it, we shall expect once again to see that it looks round, and not square, when we see it from a considerable distance, and we shall also expect that in the future it will appear square, and not round, when seen from closer at hand.

The above points seem to me to involve a fairly radical departure from most discussons of the bases on which certain appearances of objects are discounted as nonveridical; as a consequence, I would expect that readers might wish to challenge the way in which the concept of an object's "stability" has been used. For example, it might be held that this characteristic is attributed to objects simply because of past tactile experiences with them, and not on the basis of any clues given directly through vision. Such tactile experiences might be identified with the fact that an object such as a tower feels "solid" (in the sense of being nondeformable) when touched by our hands. Therefore, in so far as we trust our past experiences of touch, we would regard any object which looked like a tower as an object which possessed one and only one permanent shape. And if it is also true, as has sometimes been claimed, that we place more reliance upon touch than upon sight, it would be natural for us to accept the tower as being square since that is how it felt to us when we were close to it and when it also looked square.

However, a series of conjectures of this sort will not succeed. In the first place, it is not true that we *always* place greater reliance upon touch than upon sight. To be sure, there are many cases in which we do accept what we feel, rejecting what we see. For example, I recall an instance in which I saw a picture of a Greek low relief in an exhibition of photographs, and it looked so strikingly three-dimensional that I ran my finger over its surface to convince myself that it was both smooth and flat. Although it still looked to me as if it had a rough texture and was a low relief, I did of course disregard the evidence of my vision and rely upon touch. On the other hand, there are equally striking cases in which we trust what we see rather than what we feel. For example, when someone blindfolds us and tests our tactile sensations with calipers on the back of our hand, we may feel only one point of stimulation although there are two points at which the calipers press upon our skin; when the blindfold is removed, we have no hesitation in saying that there are in fact two points although we still feel only one. Which sense modality we trust is, then, a complex matter, and it cannot be said that under all circumstances we trust one rather than another. In the second place, even if it were true that we always gave precedence to touch over sight when what is involved is the perception of shape, this would be irrelevant to the main issue. The fact that we find that objects like towers are solid in the sense that they do not change shape when we touch them (or even when we push against them) does not provide a more decisive clue to the continuing stability of their shape than we could obtain through vision. This should be obvious. In order to reject the roundness of a tower which we see from a distance on the basis of the squareness which we feel when we approach close enough to touch it, we must assume that it did not change shape as we approached it. In other words, we must make precisely the same sort of assumption with repect to touch as we made with respect to vision: that an object such as a tower is stable with respect to its qualities.

One other possible way of escaping this conclusion may perhaps occur to the reader. It may be thought that the stability which we attribute to an object such as a tower is simply a product of the fact that in these cases the testimony of our various sense modalities is convergent. However, such an analysis would also be both mistaken and irrelevant. It would be mistaken since we do not in fact confine our judgments regarding the true characteristics of objects to cases in which the different sense modalities yield converging rather than diverging results. Examples of this are provided by cases such as those just mentioned in which, on the basis of touch, we reject what we see even though the visual illusion persists; and, conversely, cases in which what comes through vision is accepted though an error in tactile perception persists. Furthermore, even were the contention not mistaken, it could be shown to be irrelevant in most of the cases in which the stability of a material object is assumed. When we rid ourselves of contradictions by citing the converging testimony of the senses, as we do in the case of the oar in the water, we usually cite testimony which is present at two different moments of time. In such cases we are therefore forced to presuppose precisely the same sort of stability in the characteristics of the object which the convergent testimony of our senses was to have proved.

Now, it is obvious that there are many possibilities of mistakes when we attribute stability to objects and when we use this stability as the basis on which we reject some of the testimony of sense perception. The particular respects in which we regard an object as stable may in fact have been changing very slowly in ways which were imperceptible to us; or it may indeed have hitherto been stable but not continue to be so; or we may have erroneously assumed that because it was stable in some respects it would also be stable in others. Experience alone can vindicate our beliefs in such matters, and it is on such experience that we must rely. However, were we simply to *tabulate* experiences without formulating *hypotheses* to explain why in some cases stability is to be found, and in others it is apparently lacking, it is doubtful whether we could ever defend the view that any objects are in fact stable. Among the hypotheses which we would need for the sake of bringing order into experience and defending our view that objects may be stable even though their appearances vary, would be an hypothesis such as the generalization that "distance alters the appearance of objects." Such a generalization would allow us to *defend* the assumption which we do in fact make when we hold that a tower which looks round from a distance is actually square, and it would also explain whatever differences in color we might observe when we see a tower from a distance and then inspect it from close at hand. Insofar as this generalization is confirmed, it is possible to reconcile the stability which we attribute to objects with the fact that they present different appearances when seen from different distances. Thus, our direct perceptual experience and our explanatory generalizations would be mutually reinforcing. However, as one seeks to refine what it means to say that "distance alters the appearance of objects," one sees that even in ordinary experience—without the aid of scientific theories—we distinguish various types of instance. There are those cases in which distance affects vision because our sight is not sharp enough to discriminate the details of that at which we are looking; there are other cases in which distance cloaks what we are seeing because of factors such as an intervening haze. Thus, in some cases at least we might give alternative explanations of why, when we look at a distant object, we see the shape which we do; similarly, we might give different explanations of why the color of that object appears as it does. If we then follow this path of refining the generalization that "distance alters the appearance of objects," we are obviously starting on the path of scientific explanation. Along this path, a scientist will begin to inquire into the nature of light, and ultimately speak of reflection and of selective absorption, of diffusion and diffraction, and the like; he will, as we have noted, also be forced to speak of the properties of what we regard as solid material objects, but which—according to his analysis—possess properties of which, in ordinary experience, we cannot be aware. In this respect, the path pursued by the scientist seems to depart from anything in which we have an interest on the basis of perceptual experience alone. However, when we take into account the fact that the inquiries which the scientist pursues started from the same sorts of consideration which we find ourselves forced to use as a means of defending a realistic interpretation of what we perceive, a continuity between science and common sense will be recognized. This continuity is easily overlooked if we forget the history of the sciences and are only concerned with the point which has by now been reached by scientific inquiries.

We shall see that the same situation obtains with respect to the other two types of consideration with which I shall next be concerned. First, however, let us draw together what we have noted in connection with the illustration of the square tower which looks round from a distance—that is, with respect to what are frequently called "illusions." If we may take this instance as typical of some of the occasions in everyday life in which we reject various aspects of our perceptual experience, what we note is that the perception of objects involves far more than a congeries of relatively independent sensations which are supposed to be able to depict only "the surface-qualities" of objects. What we perceive also includes characteristics such as stability or lability, and it is on the basis of such character-istics that we sift and we criticize many of the more specific qualities of these objects, such as the shapes which they present to us, or the colors which they appear to have. Stability and lability are not, of course, the only such properties of material objects. For example, what in ordinary life we consider to be a material object is seen as having unity and solidity, rather than being a congeries of separable individual facets.[61] Once one abandons the assumption of a correlation between what is directly perceivable and what can be represented on our peripheral sense organs, there is no reason to deny that qualities such as stability or unity are as basic in percep-tion as are colors or shapes or the other *qualia* with which episte-mologists have been more frequently concerned. Thus, in abandon-ing what has sometimes been called atomistic sensationalism, per-ception may be recognized as providing a far richer vein of knowl-edge than it is conventionally assumed to do.[62] Furthermore, some of the perceptual qualities which are thus seen to be basic may serve

[61] In this connection I would hold that what we regard as material objects are also frequently seen by us as acting on other material objects, and as being acted upon by them. For example, what we see when we see one billiard ball strike another is a transference of force, not—as Hume would suppose—merely colored shapes in a sequential series of different positions.
Some may wish to challenge this assertion, and be inclined to side with the Humean analysis of what is given in direct sense experience. This, however, reduces to a question of either genetic or analytic psychology, with reference to which empirical investigations are directly relevant. There is perhaps no clearer example of how crucial psychology may be for epistemological analyses than that provided by Hume. This is a point on which he himself was entirely clear, as one can see both from the title and the introduction of the *Treatise*.

[62] There are of course those who would protest that I am including far more in what is "directly perceived" than is legitimate. They would wish to attribute such qualities either to "judgment" (e. g., Descartes) or to "the imagination" (e. g., Hume). In either case, such a contention seems to me plausible only if we *first* assume that what is to be included in direct sense perception is confined to what is capable of being represented on our peripheral sense organs. Since I do not regard this as an assumption which one is entitled to make, I would not feel myself to be seriously challenged by either of these positions.

to suggest ways in which the concepts employed by physicists are linked to everyday life—ways which we would never suspect if we took specific *qualia* such as size and shape and color to be all that we can experience in relation to material objects. To that point I shall later return.

I now wish to discuss a second area in which any critical sifting of sense experience puts one squarely on the path which scientific inquiry also follows, and which therefore provides another type of case in which there are important elements of continuity between our ordinary conceptions of material objects and the presumably quite different conceptions at which the sciences have currently arrived. While our first illustration of this continuity was concerned with the criticism of contradictions within sense experience, the cases which I shall now examine arise out of attempts to explain the characteristic modes of action of different sorts of material objects.

I think that it may be taken for granted that in ordinary experience there are many instances in which we explain the action of a particular sort of object by appealing to its directly observable properties. We expect that something which feels very heavy, and which we can only support with effort, will fall to the ground if we release it; similarly, we expect that something which feels solidly compact, and resists any effort on our part to break or to bend it, will not be readily broken or bent, and will not disintegrate, when it comes into contact with objects other than our own hands. Whether our beliefs in these matters are to be explained wholly in terms of past experience, or whether other factors are also involved, is not a matter which is of present concern.[63] What is important is

---

[63] It is perhaps worth noting, however, that there may be a difference between those cases in which we explain an object's action with reference to the particular *class* of natural or man-made objects to which it belongs, and those cases in which our explanation involves reference to a particular *quality* of such an object. In the former case one would naturally expect that our causal belief rests upon past observation of how this class of thing may be expected to act; in the latter, the same situation *might* obtain, or else some other genetic account might be deemed plausible. In point of fact I believe that there are some cases in which sensible qualities directly suggest modes of operation, independently of past experience; if there were not such cases it might be difficult to explain many learning processes. However, this class need

that in ordinary experience—for whatever reasons—we often take the directly sensed qualities of objects, such as their heaviness or their solidity, to be clues as to how these objects may be expected to behave.

However, even within ordinary experience we soon learn not to rely unhesitatingly upon such sensible clues. For example, we learn that there are some objects which feel no less compact and solid than do others (e. g., objects made of glass or of china), but which will nonetheless shatter if they are struck a sharp blow. Our common-sense conjectures to explain such differences may light upon various factors, depending upon the cases with which we are confronted, or with which we have been familiar in the past. Some of these conjectures might, for example relate to other sensible properties of those objects which act in unanticipated ways. For example, we might initially think that among objects which feel equally hard to our touch, some shatter because they are thin; thus we would be relating their fragility to their shape. However, in other cases there may not be any particular sensible quality which would seem to explain fragility, and we might conjecture that there was an un-noticed difference between the inner composition of an object which breaks and those objects which resemble it but do not break. In such cases the factor which we would hold responsible would not be one which had been directly observed. To be sure, neither of these two types of conjecture would arise were it not for past observations and a comparison of instances in which breakage did or did not occur. It does not follow, however, that the relevant past experience must be confined to noting that some particular sensible quality, or some set of such qualities, is uniformly connected with the fact that an object breaks. Clearly this would not have been true in the second type of case, where past experience suggested no specific correlation between breakage and any directly sensible quality, but only the very general truth that breakage may be associated with the sort of stuff out of which objects are made.[64] As be-

not constitute a large sector of those cases in which we offer causal explanations of why objects behave as they do.

[64] The attempt of philosophers such as Berkeley and Hume to reduce all statements concerning the dispositional properties of objects to statements

tween these types of conjecture (both of which go beyond an appeal
to the immediate data of how hard various objects may have felt
to our touch), I do not see that we consistently place more confidence in one than in the other. For example, I should not suppose
we would feel greater hesitation in accepting as an explanation of
breakage the fact that an object may have been made of stuff different from that of which other objects which had a similar appearance were composed, than we would in accepting the explanation
that it broke because it was too thin to withstand a sharp blow.
Thus, in everyday experience, explanations of why particular material objects behave as they do are not invariably couched in terms
of qualities which are directly perceptible in the particular instances
which we wish to explain.

This point may be pressed one step further. As our experience
widens, we shall have to depart even more radically from a reliance
upon sensible qualities alone, for it will become clear that neither
of these original conjectures will withstand scrutiny: if each is
plausible in some cases, each will also prove inadequate in others.
Consequently, if we are to put forward a generalization which will
be adequate in all familiar cases, we shall have to relate the two
sorts of conjecture to one another, attempting to state why *both*
the shape of an object and its inner composition are related to the
fact that it does break, or that it fails to break, under particular
sorts of circumstances. When we attempt to do so we shall find that,
in the end, what we say about the dispositional properties of objects,
such as their fragility, or their malleability, or their elasticity, and
the like, will involve us in theories which are not necessarily formulated in terms of directly observable qualities. Furthermore, when
we wish to confirm the accuracy of such theories we shall have to
do so by means of inference, and not through sense perception alone.
This, then, is one example of the way in which our common-sense
conjectures tend to lead us from generalizations which involve only

---

analyzable into particular sets of sensible ideas is connected, of course, with
their assumption that we cannot frame abstract general ideas. This is a further
point at which I would hold that psychological inquiry is capable of testing
(and I believe refuting) a basic epistemological assumption.

the directly observable qualities of objects to more general theories concerning the components and the structure of these objects— theories which not only serve to explain fragility, but which systematically connect it with malleability, elasticity, and the like.

When one views the rise of the modern physical sciences in the light of such considerations, it can scarcely be held that the corpuscularian natural philosophy of the seventeenth century was a metaphysical hypothesis which mistakenly departed from the observational base upon which empirical inquiries must rest. What had become clear to those who investigated natural phenomena was that transdictive inferences were necessary to account for the facts of ordinary experience. The attack of Berkeley on the legitimacy of such inferences, and Hume's attack on the possibility of justifying a realistic interpretation of them, are familiar. However, it is not always noted to what extent the crux of that attack—the analysis of causal inference proposed by Berkeley and developed by Hume—depended upon philosophic presuppositions, rather than upon an unbiased examination of the grounds of our causal beliefs. A direct examination of our causal beliefs would reveal that, in some cases, causal inferences do not in fact involve the correlation of a particular set of sensible qualities with the occurrence of a certain type of phenomenon. On the contrary, as we have just seen, both laymen and scientists often seek to explain the behavior of material objects in terms of the unobserved parts of these objects, and the actions of those parts upon each other and upon other objects. This was what underlay the corpuscularian theory, and was involved in its application to phenomena such as gravitational attraction. To be sure, after such a theory of the action of unobserved entities has been formulated as a means of explaining observed regularities in the behavior of macroscopic objects, it will be possible to reformulate that theory by using terms which refer only to directly observable regularities among these objects. However, this does not imply that one could have arrived at the theory itself if one had originally been able to assign meaning only to those terms which refer to directly sensible properties. Furthermore, it is necessary to consider whether even our best-confirmed generalizations would be regarded by us as being well-confirmed if what we took to be the relevant data were limited to those qualities of macroscopic objects which can be directly observed.

To make these points clear, let us consider the means by which the speed of light was first established, using first Roemer's observations and then the Fizeau experiment. Were either of these scientists to have been confined to the analysis of causation which was advocated by Berkeley and by Hume, the very subject matter of their inquiries could not have been said to have existed, since in the case of Roemer one could not speak of seeing *light*, but only of seeing or not seeing Jupiter and its satellites at various times. Similarly, in the Fizeau experiment, light cannot be seen as traveling from its source to the mirror and thence to the eye: all that could be seen through the serrated edge of the rapidly revolving wheel (or what one could be prevented from seeing) would be that which was taken to be an image of the source.[65] To be sure, once one had made these observa-

---

[65] That this is the case is made clear in Mach's original definitions, even though it is concealed in his accounts of the relevant experiments. In those definitions, after distinguishing between self-luminous and dark objects he defined " illumination " as follows: " We shall call the sum total of the physical relations between one object and another, determined by the feature of the visibility of the first object, the condition of *illumination*." And he thereupon said: " The mechanism *imagined* to be involved, conditioned by the first object, is designated briefly as *light* " (*Principles of Physical Optics*, p. 2). Thus, it is presumably with the conditions of the illumination of objects, and not with " light " (realistically interpreted) that the Roemer observations and the Fizeau experiment were concerned.

In order to be wholly certain that the point of the above examples is not missed, I shall now quote C. D. Broad's *résumé* of the experiment:

Fizeau's experiment.—Light is sent through a hole, in front of which is a cogwheel. When one of the teeth of the wheel is in front of the gap, light cannot pass; otherwise it can. The light travels some considerable distance, and is then reflected back along its old course, and the image is viewed from behind the cogwheel. If the passage of the light between the source and the mirror and back again be instantaneous, the image will be visible, no matter how fast the cogwheel revolves; for if *no* time has elapsed, the cogwheel cannot have moved *any* distance since the flash left it and before the light returned to it. The gap cannot, therefore, have become shut, in the meanwhile, by the rotation of the cogwheel. But if any finite time elapses between the departure and the return of the light, it must be possible to cause the original gap to be replaced by the next tooth by the time that the light returns, provided that the cogwheel has moved fast enough. In that

tions and experiments, and had deduced the speed of light from them, one could readily translate the whole story—including the speed of light, which cannot be directly observed—into a vocabulary in which only terms referring to the directly perceptible characteristics of macroscopic objects were to be found. However, such a translation could only be made because we were not originally bound by the restrictive covenants of a phenomenalistic interpretation of the world; that is, because we took light to be more than "an *imagined mechanism.*" Had one from the outset regarded it as such, Fizeau's apparatus would not have been designed, nor would the relations of succession which were observed by the use of that apparatus have indicated what we, as well as Fizeau, correctly take them to have indicated: that light travels from a source at a velocity which was measured by means of the size of the apertures and the speed of the rotation of the wheel.[66]

The import of this conclusion for a critical realism is the fact that it permits one to claim that scientific inference does give information concerning the independent existence of material objects, and concerning the properties which these objects may legitimately

case no image will be seen. If the speed of the wheel be now increased enough, the image ought again to be seen, since the wheel will have turned so far in the time taken by the passage of the light that the next gap will be in position to admit the reflected beam when it returns. It is found that the image can be made to disappear by rotating the wheel fast enough, that it can be made to reappear by rotating the wheel faster, and that the wheel needs to be rotated faster and faster the nearer the mirror is to the source, in order to make the image disappear (*Scientific Thought*, p. 377).

Roemer's observations are discussed by Broad on pp. 378–79, and more detailed accounts of the methods of both Roemer and Fizeau (as well as the other relevant experiments) are to be found in Mach's book, which is cited above.

[66] I believe—though I do not wish to press the point—that this case is strictly comparable to another case in which one must assume factors to have existed independently of our observations, but in which one can later reformulate what occurred without making use of these factors, defining them in terms of directly observable elements. The case I have in mind is that of understanding certain forms of human action. In understanding some actions we may have to take into account the agent's intentions, but once we have inferred what these intentions actually were, and once we have seen their effects, we can reformulate a description of his action as if these intentions were merely constructs used to describe his overt behavior under these circumstances and to relate it to his behavior under other circumstances.

be claimed to possess. The fact that such objects possess qualities very different from those which are directly presented to our senses should not be used as an argument for any form of general skepticism. As we have seen, in ordinary experience itself we reject some directly observed qualities because of other properties, such as stability and permanence, which we hold to be more basic characteristics of material objects. Thus, through both our ordinary experience and scientific inquiry we gradually come to build up a fuller and more detailed knowledge of the nature of the world of objects than direct sensory experience provides. A critical sifting of the appearances of objects on the basis of our experience, and the cautious use of inferences beyond what we directly experience have permitted men to reach conclusions which are, I believe, only capable of being challenged through an arbitrary philosophic fiat, not on any assignable empirical grounds.

This conclusion may now be fortified if we turn to the third point at which there is a continuity between common-sense inference and scientific inquiry. This point is to be found in considerations regarding the causal chain involved in the process of perception itself.

Were we simply to take the results of what we learn from the physical sciences concerning the nature of all material bodies and compare these results with that which we directly experience, we should truly be at a loss to know how to reconcile these two descriptions of the world. Even the conviction that one cannot escape following the course of a critical sifting of our ordinary experience, nor escape the necessity of drawing transdictive inferences, would not reconcile us to this state of affairs. Those who would seek to avoid the problem by keeping the two forms of description wholly insulated from one another will—as I have tried to show—surely fail. Fortunately, however, a way remains open by means of which we can comprehend the relations between what we directly experience and what has been learned through scientific inquiry concerning the nature of the physical world. As we shall now see, that way lies along the path which scientific inquiry has followed in attempting to understand the causal processes involved in perception itself.

As we have noted,[67] in ordinary experience we take perceiving to involve some sort of "transaction" between ourselves and that which we perceive. While we take for granted that our sense organs are somehow especially involved in that transaction, even the crudest details of what actually occurs, and how it occurs, are not usually suggested by anything of which we are directly aware. Consider, for example, the case of vision. As the history of early Western thought indicates, so far as direct experience is concerned it is surely not clear whether effluences come from the object to the eye, whether that organ itself casts light upon what we see, making it visible, or whether (as Plato held) there is a conjunction of such forces flowing from opposite directions. These contradictory conjectures could scarcely have been proposed if ordinary experience had suggested the relations which obtain among the elements which we all take to be essential to vision: an object, the eye, and that which we identify as light.

Nor is it only in the case of vision that we are at a loss to understand through direct experience just what are the basic processes involved in sense perception. Even in the case of touch, the transaction between our sensing organ and the object and its qualities is by no means clear. To be sure, *contact* is experienced as playing the major role in touch, but what actually occurs in this contact is itself by no means obvious. There may, to be sure, be some cases in which no puzzles seem to be presented. For example, an object which looks smooth will usually also feel smooth, and the correlation between what we can see and what we can touch will seem perfectly evident and be wholly expected when we compare the qualities of, say, a sphere and a cube. Yet even in such cases we would have only the crudest notion of why, upon contact, a sphere or a cube presents us with the tactile qualities which it does: on the basis of direct experience alone we could only say something vague about how the angularity of the junctures of the sides of the cube made a different sort of contact with our hands than did the continuity of the surface of the sphere.[68] The vagueness of this sort of answer becomes even

[67] Cf. above, p. 205.
[68] If, with respect to this illustration and with respect to the famous Molyneux problem of the man who has been born blind, it is said that our expecta-

more obvious if we consider some instances of thermal sensations. Within a certain range of temperatures, the warmth (or the coolness) which we feel in an object may seem to be known because it is directly transferred from the object to our skin: for example, when we feel a bath to be warm, our body will also feel warm, and a cool tub of water seems to transfer its coolness to us. Yet, the notion of a transference of a quality in such a case is both crude and unclear. For example, should one regard it as similar to what occurs when the color of a piece of chalk is transferred to our fingers; or is it like the transfer of motion; or is it of some different sort entirely? [69] The difficulty becomes even more apparent when we find that very hot objects and extremely cold objects transfer not sensations of heat or of cold, but cause a stinging pain. Thus, even apart from any questions explicitly referring to the specific nature of our cutaneous sensing organs, the most basic features of tactile perception are not really very much clarified by saying that what occurs does occur because of the contact between our bodies and the objects we touch. And what is true with respect to touch is surely paralleled by the vagueness of the notions which one would possess concerning our hearing, or the sense of smell, or taste, if one were wholly unfamiliar with the results of scientific inquiries.

An examination of the conjectures by means of which men have sought to overcome this opaqueness in the relations between the objects which we perceive and the means by which we perceive them shows that such conjectures have almost invariably involved hypotheses concerning unobservable entities existing in the physical

---

tions are wholly due to the effects of learning, that conviction would only serve to fortify my main point, and exclude the above illustration from the ranks of the exceptional cases. However, for my own part, I do not find that the empiricist theory is the most plausible of the alternatives with respect to this illustration or with respect to the problem set by Molyneux.

[69] The unclarity of the notion of transference could also be illustrated with reference to tactile qualities other than those associated with the thermal properties of objects. For example, is the same sort of " transfer of qualities " involved when we discriminate between a rough and a smooth object as is involved when we discriminate between the wet and the dry? Also, one can raise the more general question of why in some cases mere contact is sufficient for tactile sensations of specific sorts, while in other cases the contact must also involve motion.

world.[70] This is not surprising, for even if men had known all that there was to be known about the anatomy of the sensing organs, it would still have been necessary to formulate hypotheses about the nature of material objects and (in some cases) about the medium existing between such objects and our organs, in order that the transaction involved in perceiving could be explained. Consequently, it should occasion no surprise that the way was not open for a continuous development within the theory of sense perception until Western thought had obtained a generally accurate view of the constituent elements present in all objects which are capable of being perceived, and a knowledge of how these elements could affect other objects (including our organs) over an intervening distance. In short, it is no accident that the foundations of an adequate experimental science of sense perception developed after the rise of the modern physical sciences.

However, the physical sciences alone would have been incapable of providing an adequate theory of perception: first anatomy and then psychophysics and finally neurophysiology have had to develop in order for us to begin to be in a position to understand how what we directly perceive is related to what the physical sciences can tell us about the nature of material objects. Crude as our knowledge remains, it is no longer correct to hold that a knowledge of neurophysiology cannot possibly give us insight into how the qualities of that which we experience are related to what occurs in the brain. When, for example, we perceive one object as larger than another, or when we perceive two objects as being first close to one another and then farther apart, an account of the correlated brain events presumably explains why we perceive these characteristics under these particular circumstances.[71] It is to be noted that brain events

---

[70] Perhaps the best accounts of the theories held by Greek and Medieval figures are to be found in Hermann Siebeck, *Geschichte der Psychologie.* Parts I, 1 and I, 2 were published in Gotha in 1880 and in 1884. The work as a whole was apparently left unfinished, but supplementary studies of late Medieval figures are to be found in volumes I, II, and III of the *Archiv für Geschichte der Philosophie* (1888–1890) and in *Beiträge zur Entstehungsgeschichte der neueren Psychologie* (1891).

[71] I do not wish to take issue with the Leibnizean challenge that even if

are also connected with the nature of stimuli affecting our sense organs, and that these stimuli, in turn, had their sources in the physical objects perceived by us. Therefore it follows that the neurophysiological account will (in some cases) help to provide a connecting link between that which we perceive and the physical characteristics of the object perceived. Thus, it should not be considered either an accident or an inexplicable mystery that we, in direct experience, arrive at discriminations which resemble those made by a physicist when he distinguishes among objects in terms of their sizes, their shapes, their distances from one another, and the like. Furthermore, even when the specific neurophysiological correlates of various qualities present in direct experience are not known, the well-authenticated general hypothesis that there are such correlates makes it inescapable for us to say that a physicist's inquiries in, say, optics will be relevant to the fact that we distinguish among objects in terms of their colors, or that the visual appearance of rainbows is different from what we see when we look at a solid and pigmented surface. Thus, the physiological processes involved in sense perception form a mediating link between that which occurs in the physical world and that which we directly experience through sense perception.

When the physiological factors involved in perception are placed in the context in which they have just been discussed, some of the traditional hostility to an emphasis upon this process should be overcome. If I am not mistaken, this hostility springs in large measure from the fact that if we relate what is perceived to a physiological account of what occurs in our nervous systems, it would seem as if all perception were inescapably " subjective," occurring

---

the brain were as large as a mill, we could not learn by inspecting it that the operation of its parts would give rise to consciousness (cf. *Monadology*, # 17, and the passage from *Commentario de Anima Brutorum*, quoted by Latta in his edition of the *Monadology*, p. 228). The connection between consciousness and the nature and functioning of the nervous system must, in my opinion, be accepted as " brute fact." However, this does not imply that, once we acknowledge this connection, we cannot correlate what occurs within direct experience with characteristics of these physical processes, and that these correlations will not serve to explain why particular aspects of our experience have the relations to one another which they in fact have.

as it does simply because of what happens within us—and, more particularly, in the secret compartments of our own brains. To such a view, thus phrased, one can—I am sure—understand our natural hostility. However, when this physiological process is viewed as merely one link in a sequence starting from a physical object, and dependent upon characteristics possessed by that object, it no longer seems to pose the paradoxes of a complete subjectivity and the mysteries of how we can "project" that which we see.[72] Instead, it becomes a way of explicating how it is that we can come to have an awareness of objects which exist independently of us. To be sure, the characteristics of these objects should not be assumed to be identical with what, in direct experience, we take them to be. Yet that is a conclusion which is forced upon us by what we know of the nature of physical objects, and how they are to affect the human organism: it is not an implication of psychophysical considerations taken by themselves.

That epistemological discussions cannot avoid introducing considerations concerning the physiological factors involved in perception may be made plain in another way, without approaching the matter through any questions which are in the first instance related to the sophisticated experimental sciences. Consider the fact that in our ordinary experience we are frequently forced to the conclusion that the question of which qualities of objects we are able to perceive depends upon the nature and the condition of our sense organs. This becomes obvious when we note that (for example) we can sometimes distinguish what we cannot always distinguish, not because of changes in the nature of the objects themselves, but because of the state of our health, or the fact that we put on glasses, or for any other of a number of reasons which are related to ourselves. And comparisons between what different persons are able to perceive under the same conditions clearly demonstrate for us that in the transaction known as perceiving much depends upon the nature of

[72] It was Ewald Hering who first proposed the solution to the problem of why what we perceive is localized as being outside of us even though it is dependent upon what occurs in the brain. Ernst Mach followed his suggestion in *The Analysis of Sensations*, but the fullest discussion of it is to be found in Köhler, *The Place of Value in a World of Facts*, pp. 127-41.

our own bodies, and this dependence is *not* a matter of which we are directly aware. This conclusion, well known to common sense, immediately puts us on the path to physiological explanations of what occurs in sense perception—as soon as we have the knowledge which must be presupposed in order successfully to carry out such investigations. And this path, as we have seen, leads directly to psychophysical and neurophysiological inquiries; and, finally the whole chain of causation which is involved permits us to comprehend how that which we perceive is related to what exists independently of us. Therefore, no matter how startling the implications of physiological considerations—when taken by themselves—may appear to be, they serve to make veridical perception more intelligible than it would otherwise be. Thus they tend to complete our understanding of what we ordinarily assume, but concerning which direct experience fails to yield adequate information: that perception is a transaction in which what we perceive depends upon the nature of that with which we are in contact through our sensing organs.

I have now concluded my brief discussion in which three different topics were shown to lead from familiar facts concerning sense perception into further, more systematic inquiries which tend either to supplement or to alter our common-sense views of the nature of the physical world. The first of these was found in the manner in which we are forced to pass from what appear to us to be contradictions within sense experience to more general notions of what characteristics are basic in material objects, as well as to explanations of how various factors (such as distance) tend to influence what we perceive. On the basis of such considerations it became clear that one traditional epistemological doctrine had to be rejected: the doctrine that we do rely, and must rely, upon what can be directly present to the senses for any knowledge that we may have of the nature of the physical world. The second topic showed that our common-sense attempts to explain the behavior of material objects inevitably lead to scientific explanations in which the actual nature of these objects is shown to be different from what direct perception would lead us to believe it to be. Finally, in the third instance it was shown that a common-sense conception of what is necessary if we are to speak of perceiving an object involves a causal chain of processes, of whose individual steps we are usually not directly aware. Not only does this account fill in the details of what would otherwise be an unintelligible set of relations between ourselves and what we perceive, but it also provides a clue as to why—in spite of the radical qualitative differences between them—there is a congruence between common sense and scientific descriptions of the nature of the physical world.

When these three topics are taken into account, it would, I believe, be unwarranted to say that what we perceive is to be identified with that which exists independently of perception; but it would be equally unwarranted to hold that we do not, because of this fact, have any concrete knowledge of the nature of what exists independently of our perceiving it. This knowledge, as the foregoing discussion should have served to make clear, has been acquired *in ambulando*: it has been implicit wherever common sense has found good grounds for rejecting some perceived qualities in favor of attributing other qualities to those objects, and it has become explicit through scientific inquiries. Were this explicit knowledge to be challenged on epistemological grounds, a justification of it could be given through tracing the phenomenological and scientific steps on the basis of which it had finally been reached. That this would provide adequate justification seems to me clear: for it surely cannot be contended that the general methods which have marked the development of modern science are different from what in ordinary experience we would regard as providing a justification of belief. In fact, it probably does not go too far to say that the methods actually used in the sifting of scientific evidence provide as clear a model of what constitutes justified reasoning as can be found. To trace the essential phenomenological and inferential steps by means of which, using this reasoning, our modern conception of the world has come to be what it is, remains one of the most comprehensive and challenging tasks for historians of science. If the fundamental conviction of this book is correct, such a task would not only be of historical but of epistemological significance.

APPENDIX

The reader who knows the epistemological theories of Wolfgang Köhler will readily recognize how closely my views are connected with his. It is scarcely necessary for me to state that the debt is wholly my own.

Were I to write a systematic treatise on epistemology, I should be inclined to adopt his terminology in many respects, particularly his distinction between "material objects" and "physical objects," and between "body" and "organism." Furthermore, I would surely lay emphasis on a dual use of the concept of "isomorphism," as does he. The latter concept seems to me important regardless of what may ultimately be the fate of Köhler's neurophysiological hypotheses concerning the factors underlying psychophysical isomorphism. To be sure, those hypotheses are not unconnected with the range of phenomena to which we may expect the isomorphic relationship to apply; however, Köhler's own experiments in the field of perception would seem to supply sufficient evidence for the existence of an isomorphic relationship, regardless of what neurophysiological hypotheses must be employed in order to explain it.

There is, I believe, one fundamental point at which my views may be markedly different from those of Köhler, and that is with respect to my use of the historic development of scientific inquiry as being relevant to epistemological questions. Had I been satisfied with his argument for "transcendence," as put forward in *The Place of Value*, I might not have laid the stress which I have done on the piecemeal, self-corrective development of transdictive inference in the history of thought. It is because I have wished to stress the epistemological significance of this process that I have not made more use of Köhler's terminology and his arguments than I have in fact done.

BIBLIOGRAPHY

Only titles actually cited in the preceding studies are included in this bibliography. There are, of course, important works relevant to each of these studies to which no explicit reference has been made.

AARON, R. I. *John Locke.* (2nd ed.) Oxford, 1955.
——— and GIBB, J. *An Early Draft of Locke's Essay, Together with Excerpts from his Journals.* Oxford, 1936.
ANDERSON, F. H. "The Influence of Contemporary Science on Locke's Methods and Results," *University of Toronto Studies in Philosophy*, II (1923), No. 1.
AUSTIN, J. L. *Sense and Sensibilia.* Oxford, 1962.
AYER, A. J. *The Problem of Knowledge.* London and New York, 1956.
BACON, FRANCIS. *Works.* Collected and edited by J. Spedding, R. L. Ellis, and D. D. Heath. 14 vols. London, 1862–76.
BAEUMKER, C. "Ueber die Lockesche Lehre von den primären und sekondären Qualitäten," *Philosophisches Jahrbuch*, XXI (1908), 293–313.
BARKER, S. F. *Induction and Hypothesis.* Ithaca, 1957.
BERKELEY, GEORGE. *The Works of George Berkeley, Bishop of Cloyne.* Edited by A. A. Luce and T. E. Jessop. 9 vols. London, 1948–57.
BLAKE, R. M. "Sir Isaac Newton's Philosophy of Scientific Method," *Philosophical Review*, XLII (1933), 453–86. [Reprinted in Blake, Ducasse, and Madden, *Theories of Scientific Method: The Renaissance through the Nineteenth Century.* Seattle, 1960.]
BLOCH, L. *La philosophie de Newton.* Paris, 1908.
BOAS, MARIE. "The Establishment of the Mechanical Philosophy," *Osiris*, X (1952), 412–541.

――――. *Robert Boyle and Seventeenth Century Chemistry*. Cambridge [Eng.], 1958.

―――― and RUPERT HALL. "Newton's 'Mechanical Principles'," *Journal of the History of Ideas*, XX (1959), 167–78.

――――. For another joint article, cf. entry under A. Rupert Hall.

BOURNE, H. R. Fox. *The Life of John Locke*. 2 vols. New York, 1876.

BOYLE, ROBERT. *The Works of the Hon. Robert Boyle*. 6 vols. London, 1772.

――――. "Unpublished Boyle Papers": cf. Westfall, as cited below.

BRAITHWAITE, R. B. *Scientific Explanation*. Cambridge [Eng.], 1953.

BRETT, G. S. *History of Psychology*. 3 vols. London, 1912–21.

BROAD, C. D. *Perception, Physics and Reality*. Cambridge [Eng.], 1914.

――――. *Scientific Thought*. London, 1923.

BURTT, E. A. *Metaphysical Foundations of Modern Physical Science*. New York, 1927.

CAJORI, FLORIAN (ed.). *Sir Isaac Newton's Mathematical Principles* cf. entry under Newton, below.

CHARLETON WALTER. *Physiologia Epicuro-Gassendo-Charltoniana, or a Fabrick of Science Natural upon the Hypothesis of Atoms*. London, 1654.

CHISHOLM, RODERICK M. *Perceiving: A Philosophic Study*. Ithaca, 1957.

COHEN, I BERNARD. *Franklin and Newton*. Philadelphia, 1956.

COWAN, D. L. "Comments on Dr. Romanell's Article on Locke and Sydenham," *Bulletin of the History of Medicine*, XXXIII (1959), 173–80.

CROMBIE, A. C. "Newton's Conception of Scientific Method," *Bulletin of the Institute of Physics*, VIII (1957), 350–62.

DESCARTES, RENÉ. *The Philosophical Works of Descartes*. Translated by E. S. Haldane and G. R. T. Ross. 2 vols. Cambridge [Eng.], 1911–12.

EDDINGTON, A. S. *The Nature of the Physical World*. Cambridge [Eng.], 1929.

FRASER, A. C. *Locke*. Edinburgh and London, 1890.

―――― (ed.). *An Essay Concerning Human Understanding*: cf. entry under Locke, below.

FREIND, J. "A vindication of his Chymical Lectures . . . ," in *Philo-*

sophical Transactions of the Royal Society of London, XXVII, No. 331, p. 330. [Abridged translated version in *Philosophical Transactions, From the Year 1700 to the Year 1720.* Abridged by Henry Jones, London, 1749, V, Part I, 428.]

GASTRELL, F. *The Certainty and Necessity of Religion in General.* London, 1697.

GIBSON, JAMES. *Locke's Theory of Knowledge and its Historical Relations.* Cambridge [Eng.], 1917.

GIUA, M. *Storia delle Scienze ed Epistomologia.* Torino, 1945.

GIVNER, D. A. "Scientific Preconceptions in Locke's Philosophy of Language," *Journal of the History of Ideas,* XXIII (1962), 340–54.

GLANVILL, J. *Plus ultra, or The progress and advancement of knowledge since the days of Aristotle.* London, 1668.

———. *Essays on Several Important Subjects in Philosophy and Religion.* London, 1675.

GREEN, T. H. "Introduction to Hume's 'Treatise of Human Nature'," in *Works,* Vol. I, Edited by R. L. Nettleship. London, 1885.

GRICE, H. P. "The Causal Theory of Perception," *Aristotelian Society, Supplementary Volume,* XXXV (1961), 121–52.

HALL, A. RUPERT and MARIE BOAS HALL. "Newton's Theory of Matter," *Isis,* LI (1960), 131–44.

———. For another joint article, cf. entry under Marie Boas.

HAMLYN, D. W. *The Psychology of Perception.* London and New York, 1957.

HUME, DAVID. *An Enquiry Concerning the Human Understanding, and An Enquiry Concerning the Principles of Morals.* Reprinted from the edition of 1777; edited by L. A. Selby-Bigge. Oxford, 1894.

———. *History of England.* 8 vols. London, 1812.

———. *A Treatise of Human Nature.* Reprinted from the original edition in three volumes; edited by L. A. Selby-Bigge. Oxford, 1896.

JACKSON, REGINALD. "Locke's Distinction between Primary and Secondary Qualities," *Mind,* XXXVIII (1929), 56–76.

———. "Locke's Version of the Representative Doctrine of Perception," *Mind,* XXXIX (1930), 1–25.

JAMES, WILLIAM. *Principles of Psychology.* 2 vols. New York, 1890.

JOAD, C. E. M. "The Status of Sense Perception in Relation to

Scientific Knowledge," *Proceedings of the Seventh International Congress of Philosophy.* Oxford, 1930.

JOURDAIN, P. E. M. "Newton's Hypotheses of Ether and Gravitation," *The Monist,* XXV (1915), 79–106; 234–54; 418–40.

KATZ, DAVID. *The World of Colours.* London, 1935.

KING, LORD. *The Life of John Locke, with Extracts from his Correspondence.* 2 vols. London, 1830.

KNEALE, W. *Probability and Induction.* Oxford, 1949.

KÖHLER, WOLFGANG. *The Place of Value in a World of Facts.* New York, 1938.

KOYRÉ, A. "Pour une édition critique des œuvres de Newton," *Revue d'histoire des sciences,* VIII (1955), 19–37.

———. "L'hypothèse et l'expérience chez Newton," *Bulletin de la Société française de Philosophie,* L (1956), 59–89.

———. "Les Regulae Philosophandi," *Archives internationales d'Histoire des Sciences,* XIII (1960), 3–14.

KUHN, T. S. "Robert Boyle and Structural Chemistry in the Seventeenth Century," *Isis,* XLIII (1952), 12–36.

LAMPRECHT, STERLING P. "Empiricism and Epistemology in David Hume," *Studies in the History of Ideas,* II, 219–52. Edited by the Department of Philosophy, Columbia University. New York, 1918.

LASHLEY, K. S. and RUSSELL, J. T. "The Mechanism of Vision. XI. A Preliminary Test of Innate Organization," *Journal of Genetic Psychology,* XLV (1934), 136–44.

LASSWITZ, K. *Geschichte der Atomistik vom Mittelalter bis Newton.* 2 vols. Hamburg and Leipzig, 1890.

LATTA, ROBERT (ed.). *Leibniz, The Monadology:* cf. entry under Leibniz, below.

LEIBNIZ, G. W. *The Monadology, and Other Philosophical Writings.* Translated and edited by R. Latta. Oxford, 1898.

———. *New Essays Concerning Human Understanding.* Translated and edited by A. G. Langley. LaSalle [Ill.], 1949.

LOCKE, JOHN. *An Essay Concerning Human Understanding.* Edited by A. C. Fraser. 2 vols. Oxford, 1894.

   Draft A of the Essay: cf. Aaron and Gibb, as cited above.

   Draft B of the Essay: cf. Rand, as cited below.

   Draft C of the Essay: cf. Aaron, *John Locke* (2nd ed.), pp. 55–73.

———. *The Works of John Locke.* 10 vols. (New edition, corrected.) London, 1823.

———. "De Arte Medica." A recent re-editing of this fragment is to be found in Alexander G. Gibson, *The Physician's Art,* pp. 13–26. Oxford, 1933.

———. *Notebooks and Journals*: cf. Aaron and Gibb, as cited above; also, Lough, as cited below.

LOEMKER, L. E. "Boyle and Leibniz," *Journal of the History of Ideas,* XVI (1955), 22–43.

LOUGH, J. *Locke's Travels in France, 1675–1679, as related in his journals, correspondence, and other papers.* Cambridge [Eng.], 1953.

LOVEJOY, ARTHUR O. *The Revolt Against Dualism.* New York, 1930.

MACH, ERNST. *The Analysis of Sensations.* Chicago and London, 1914.

———. *Principles of Physical Optics.* London, 1926.

MACLAURIN, COLIN. *An Account of Sir Isaac Newton's Philosophical Discoveries.* London, 1748.

MANDELBAUM, MAURICE. "Professor Ryle and Psychology," *Philosophical Review,* LXVII (1958), 522–30.

MEIER, JOHANN. *Robert Boyles Naturphilosophie.* Fulda, 1907.

MICHOTTE, A. "A propos de la permanence phénoménale," *Acta Psychologica,* VII (1941), 298–322.

———. "Le caractère de 'réalité' des projections cinématographiques," *Revue internationale de Filmologie,* Cahier # 1. Paris, 1948.

MOLYNEUX, WILLIAM. *Dioptrica Nova.* London, 1692.

MOORE, G. E. *Commonplace Book 1919–1953.* Edited by Casimir Lewy. London and New York, 1962.

———. *Philosophical Studies.* London and New York, 1922.

———. *Some Main Problems of Philosophy.* London and New York, 1953.

———. *The Philosophy of G. E. Moore.* Edited by Paul A. Schilpp ("Library of Living Philosophers," Vol. IV). New York, 1952.

NEWTON, ISAAC. *Sir Isaac Newton's Mathematical Principles of Natural Philosophy, and his System of the World.* Translated by Andrew Motte in 1729; translations revised, etc. by Florian

Cajori. Berkeley, 1934. [Referred to in the text as Newton's *Principia*.]

————. *Opticks*. With prefatory material by A. Einstein and E. Whittaker. London, 1931. [Reprinted with a further preface by I Bernard Cohen, New York, 1952.]

————. *Isaac Newton's Papers and Letters on Natural Philosophy*. Edited by I Bernard Cohen, Cambridge [Mass.], 1958.

————. *Unpublished Scientific Papers of Isaac Newton*. Translated and edited by A. Rupert Hall and Marie Boas Hall, Cambridge [Eng.], 1962.

————. *The Correspondence of Sir Isaac Newton and Professor Cotes*. Edited by J. Edleston. London, 1850.

————. *The Correspondence of Isaac Newton*. Edited by H. W. Turnbull. Vol. I– , Cambridge [Eng.], 1959–

O'CONNOR, D. J. *John Locke*. London, 1952.

OGLE, KENNETH N. "Theory of Stereoscopic Vision," in *Psychology: A Study of Science*. Edited by Sigmund Koch. Vol. I, New York, 1959.

OLLIAN, H. *La philosophie générale de Locke*. Paris, 1908.

PASSMORE, JOHN A. *Hume's Intentions*. Cambridge [Eng.], 1952.

PEMBERTON, HENRY. *A View of Sir Isaac Newton's Philosophy*. London, 1728.

PRICE, H. H. *Hume's Theory of the External World*. Oxford, 1940.

————. *Perception*. London, 1937.

PRICHARD, HAROLD A. *Knowledge and Perception*. Oxford, 1950.

RAND, BENJAMIN (ed.). *An essay concerning the understanding, knowledge, opinion, and assent* [by John Locke]. Cambridge [Mass.], 1931.

RANDALL, J. H., JR. "Newton's Natural Philosophy," in *Philosophical Essays in Honor of Edgar A. Singer*. Edited by F. P. Clarke and M. C. Nahm. Philadelphia, 1942.

ROMANELL, P. "Locke and Sydenham: A Fragment on Smallpox (1670)," *Bulletin of the History of Medicine*, XXXII (1958), 293–321.

————. "Grant No. 2227. Connections between Locke the Philosopher and Locke the Physician," *Year Book of the American Philosophical Society*, 1958, pp. 347–49.

ROSENBERGER, F. *Isaac Newton und seine physikalischen Prinzipien*. Leipzig, 1895.

ROYAL SOCIETY OF LONDON. *The Record of the Royal Society of London.* (3rd ed.) London, 1912.

RUSSELL, BERTRAND. *Human Knowledge, Its Scope and Limits.* New York, 1948.

———. *Problems of Philosophy.* New York, 1912.

RYLE, GILBERT. *The Concept of Mind.* London, 1949.

———. *Dilemmas.* Cambridge [Eng.], 1954.

SCHRÖDER, W. *John Locke und die mechanische Naturauffassung.* Erlangen, 1915.

SIEBECK, HERMANN. *Beiträge zur Entstehungsgeschichte der neueren Psychologie.* Giessen, 1891.

———. *Geschichte der Psychologie.* Gotha, 1880–84.

STEBBING, L. S. *Philosophy and the Physicists.* London, 1937.

STOUT, G. F. *Analytic Psychology.* 2 vols. London and New York, 1896.

———. *A Manual of Psychology.* (3rd ed.) London, 1904.

STRANG, COLIN. "The Perception of Heat," *Proceedings of the Aristotelian Society,* LXI (1961), 239–52.

STRONG, E. W. "Hypotheses Non Fingo," in *Men and Moments in the History of Science.* Edited by Herbert M. Evans. Seattle, 1959.

SYDENHAM, THOMAS. "Medical Observations Concerning the History and Cure of Acute Diseases," in *Works,* Vol. I. Edited by R. G. Latham. London, 1848.

THOMPSON, S. M. *Study of Locke's Theory of Ideas.* Monmouth [Ill.], 1934.

TOULMIN, STEPHEN E. *The Philosophy of Science.* London and New York, 1953.

WARD, JAMES. *Psychological Principles.* Cambridge [Eng.], 1919.

———. "Psychology," *Encyclopedia Britannica.* (9th ed.). 1885.

WESTFALL, R. S. "Unpublished Boyle Papers Relating to Scientific Method," *Annals of Science,* XII (1956), 63–73 and 103–17.

———. "The Foundations of Newton's Philosophy of Nature," *British Journal for the History of Science,* I (1962), 171–82.

WHITEHEAD, ALFRED NORTH. *Science and the Modern World.* New York, 1926.

WIENER, P. P. "The Experimental Philosophy of Robert Boyle," *Philosophical Review,* XLI (1932), 594–609.

WRIGHT, GEORG H. VON. *The Logical Problem of Induction.* (2nd ed.) Oxford, 1957.

YOLTON, J. W. "Locke's Unpublished Marginal Replies to John Sargent," *Journal of the History of Ideas*, XII (1951), 528–59.
————. *John Locke and the Way of Ideas*. Oxford, 1956.
YOST, R. M., JR. "Locke's Rejection of Hypotheses about Sub-Microscopic Events," *Journal of the History of Ideas*, XII (1951), 111–30.

INDEX

Aaron, R. I.: on Locke and Gassendi, 4n.; on primary qualities in Locke, 16

Abstract ideas: Locke on, contrasted with Berkeley and Hume, 116; Berkeley and Hume on, 116, 124n., 139n., 233n.

Alchemists: Locke on, 59n.; Boyle on "Spagyrists," 95

Anderson, F. H.: on Locke, 4n.

Aristotelians. See Schoolmen

Atomism:
in Locke, 1–15, 54, 60, 88
in Boyle, 1, 60, 88–89, 95–101, 103, 105
in Newton, 1, 11n., 60, 66–68, 79, 84f., 85n., 88, 88n.
and sense perception, 15
and microscopes, 66, 106, 109
and Charleton, 103n.
and scientific explanation, 235

Audition: and vision, compared, 180–81

Austin, J. L.: mentioned, 183n., 226n.

Ayer, A. J.: on phenomenalism, 120n.

Bacon, Francis: use of term "histories," 96n.; mentioned, 51, 92, 93n., 96n., 106, 219n.

Baeumker, C.: on Locke, 4n.

Barker, S. F.: inductive inference and realism, 63n.

Berkeley:
challenge to previous epistemology, 3
philosophic critique of science, 3, 116

on abstract ideas, 116, 124n., 139n., 233n.
escape from subjectivism, 121
on visual and tactile correlations, 126
use of term "philosophy," 136n.
his theory of vision and Hume, 140–43
presuppositions of his theory of vision, 143–44
on perception of distance, 150n., 224n.
ideas as coming through the senses, 164n., 169n., 224n.
transdiction rejected, 235
mentioned, vii, viii, 2, 4n., 14, 20n., 27f., 33, 60, 100n., 155n., 161, 192, 215n.

Bernier, François: Locke's contact with, 10n.

Blake, R. M.: on Newton, 80n.

Blanshard, Brand: mentioned, 179n.

Bloch, Léon: on Newton, 86

Boas, Marie. See Hall, Marie Boas

Boyle:
and atomism, 1, 60, 88–89, 95–101, 103, 105
personal relations with Locke, 3n.
on atomism and medicine, 7n.
and "histories," 8, 96n.
on primary and secondary qualities, 19, 21n., 99–101, 115
on essences, compared to Locke, 42n.
on the Royal Society, 50
on explanation, 53, 96
and alchemists, 59n., 95

255

Index

Index 259

and causal theory of perception,
1–2, 4f., 12n. f., 27, 35, 37
personal relations to Boyle, 3n.
and Gassendi, 4n., 9
medical fragments and atomism, 6–8
and "histories," 8, 96n.
and Descartes, 9, 32, 59n.
and dogmatism, 11–12
meaning of "knowledge," 12, 52, 59–60
and the Schoolmen, 12n., 32, 43–44, 59n.
on essences, 12n., 13n., 16, 41–46, 51–55, 58
acceptance of results of the sciences, 13–15
not concerned with scientific method, 13n., 52n.
origin of knowledge and atomism, 15
primary and secondary qualities, 15, 16–30, 101n.
on substance and substances, 15f., 31–46
ideas and qualities distinguished, 16, 27, 30, 34–35, 36–37
gap between his discussion of primary qualities and of substance, 29–30
relation of powers to knowledge of objects, 37, 55–58, 115n.
atomism and doctrine of substance, 37 39
on species, 43–44
and substantial forms, 44, 55
purpose of the *Essay*, 46–51
on system-building, 47–48, 59n.
and the Baconian tradition, 51
distinction between scientists and "mankind in general," 51–52, 57–58
and knowledge of real essences, 51–55, 58
on extent of knowledge, 59–60
and alchemists, 59n.
and "hypotheses," 59n.

views compared with Newton's, 85n., 87–88
on explanation in terms of motion, 99n.
motion not a necessary property of matter, 100n.
and abstract general ideas, 116, 124n.
as critical realist, 119
on sense organs in perception, 138, 224n.
compared with Hume, on external world, 150n.
mentioned, vii, 129n., 155n., 161, 171, 195, 208, 223n.
Loemker, L. E.: on Boyle, 89n.
Lovejoy, A. O.: speed of light and epistemology, 179n.
Luce, A. A.: on Berkeley, 141n., 164n.
Lucretius: mentioned, 103n.

Mach, Ernst: assumptions regarding perception of objects, 225n.; analysis of experiments on light, 236n.; on external localization of percepts, 243n.
Maclaurin, Colin: on Newton, 53n.
Malebranche: mentioned, 140
Meier, Johann: on Boyle, 89n.
Michotte, A.: on "reality-character," 153n.
Microscopes: and atomism, 66, 106, 109
Mill, J. S.: assumptions regarding perception of objects, 225n.
Molyneux, William: and Berkeley's theory of vision, 141n. See also Molyneux problem
Molyneux problem: Locke states the problem, 126n.; views of Locke, Berkeley, Hume, and Leibniz on, 126n.; mentioned, 239n.
Moore, G. E.:
his choice of examples, 172, 191, 193
does not discuss primary and secondary qualities, 174